Fundamentals of Data Science

Part III: Machine Learning

Jared M. Maruskin

Fundamentals of Data Science

Part III: Machine Learning

Cayenne Canyon Press

Dr. Jared M. Maruskin

Published by Cayenne Canyon Press
San José, California

ISBN: 978-1-941043-13-4 (softcover)

10 9 8 7 6 5 4 3 2 1

More human than a human is our motto.
—Blade Runner

Preface

The goal of this project is to lay out a mathematical foundation of data science, from basic statistics through machine learning. In my experience, many data science books largely focus on data-science applications and programming packages, and not the nitty gritty of the underlying mathematical theory, which is largely buried across numerous graduate texts in statistics and machine learning. In addition to laying out the mathematics, I also complement the reading with numerous examples in Python, and strive (especially as this series progresses) to follow an object-oriented approach to programming.

Due to the encyclopedic scope of this project, I made certain necessary trade offs in order to make a condensed version of the topics presented throughout these pages, while covering as much material as possible in a reasonable amount of time. In particular, I favored clear explanation of the mathematical principles, supplemented with working Python code, over a parade of interesting examples, which I find to be commonly available in many other wonderful texts. Most chapters are themselves condensed versions of entire textbooks, which I refer to for additional discussion and depth throughout.

This work is intended for students of data science with a prerequisite of mathematical maturity, PhDs from the broader panoply of mathematical sciences with a weak statistics background who are interested into transitioning into the field, and as a reference source for practicing data scientists. I personally belonged to this second camp: my own PhD was in dynamical systems theory, with applications to aerospace engineering (tracking space debris; see Maruskin, *et al.* [2012]) and physics (geometric theory of the variational principles of nonholonomic systems; see Maruskin [2018]), whereas my knowledge of statistics, modeling, and machine learning, on the other hand, I picked up on the job after transitioning into the data science field. Since my understanding of many of these topics was superficial or tangential, I found myself one day wishing that I had a masters degree in statistics and machine learning and figured, *no worries, I used to be a professor, I can do that myself.* So I sat down and outlined my own personal curriculum and divided it among three volumes—statistics and experiment, statistical modeling, and machine learning—which form the basis of this series. In short: my goal was to lay a foundation of expertise in statistical modeling and machine learning for myself, while leaving a foundational path for others to follow.

I am very excited to present the third and final volume of the series, which covers an introduction to machine learning. In particular, we discuss k-means clustering, tree-based methods, including random forests and XG-Boost, deep learning, and, finally, reinforcement learning. The emphasis of this third part of the series is on algorithms and coding implementations as compared to the earlier parts, which focused more heavily on statistics. This project has been a labor of love, and I've dedicated many weekends and early mornings to read, learn, solve and write the content of these pages. I hope the result benefits your learning and understanding and proficiency of the topics as it has mine.

Coding examples throughout are written in Python. I have found that writing simple simulations is an effective tool in learning theory. It also serves as a tool to validate our theoretical formulas. Several times in the process of writing this text, I developed a formula, ran a simulation, and found that it didn't work the way I expected. This led me to discover my errors and, when the simulation finally worked, served to validate that I had arrived at the correct conclusions.

Python Distribution

I recommend obtaining the free Anaconda Python 3 distribution, available from `anaconda.com`. It comes with all of the scientific packages preinstalled. Anaconda is a bundle of applications, so you can use it to launch R-Studio, Jupyter Notebook, or Spyder. I write my code in Spyder. The Sypder interface gives you access to a code editor, an iPython console, and help docs, all in one screen. I find it extremely useful running code in the console while I develop source code in the editor. I have also have several friends who use PyCharm from `jetbrains.com` as an alternative.

Many of the examples throughout this text will make use of various packages. The import statements that are used throughout are collected in Code Block 1. I will further assume the reader to have familiarity with Python. For those starting out, however, I recommend McKinney [2017] as an excellent place to start. For additional references on Python for machine learning, featuring sci-kit learn and tensorflow, see Gèron [2019].

San José, California   *Jared M. Maruskin*
                                                               April 2022

```python
import numpy as np
import pandas as pd
import sklearn, scipy
import time, datetime
import matplotlib.pyplot as plt
import pymc3 as pm
from mpl_toolkits.mplot3d import Axes3D # For 3D
from pandas import DataFrame, Series
from abc import ABC, abstractmethod
from queue import Queue
from collections import deque
from collections import Counter
from scipy.special import comb
from scipy.special import beta as BETA
from scipy.special import gamma as GAMMA
from sklearn.preprocessing import OrdinalEncoder, OneHotEncoder,
    KBinsDiscretizer, KBinsDiscretizer, StandardScaler
from sklearn.impute import SimpleImputer
from sklearn.pipeline import Pipeline
from sklearn.compose import ColumnTransformer
from sklearn.model_selection import train_test_split
from sklearn.metrics import r2_score, roc_curve, auc,
    confusion_matrix
from sklearn.calibration import calibration_curve
from statsmodels.tsa.stattools import acf, pacf
from tqdm import tqdm
```

Code Block 1: Common package imports.

Contents

Part III Machine Learning

6

Modeling Runway

In this chapter, we lay some of the groundwork for the models we will discuss throughout the rest of this book. We begin with a discussion of basic nomenclature, and then discuss common approaches for model validation. We then discuss aspects of preprocessing that are common to most data science applications. We conclude the chapter with a discussion on engineering aspects of data science; in particular, we discuss how to take an object-oriented approach to data-science projects.

6.1 Modeling and Validation

In this section we lay out some key definitions of modeling and machine learning. We then describe various metrics and methods for model validation that will be used extensively throughout the remainder of this text.

6.1.1 Models and Model Learning

We begin by laying some terminology for modeling and machine learning.

Definition 6.1. *A* model *is a mathematical representation of a system or a process.*

A predictive model *is a model for an output (or* target*) variable as it depends on a prescribed set of inputs (or* features*).*

A statistical model *is a predictive model whose output is a probability distribution for the target variable.*

Statistical models are often expressed in the form

$$Y = f(X) + \epsilon,$$

where f is a predictive model and ϵ is an *error term*, with $\mathbb{E}[\epsilon] = 0$, so that $\mathbb{E}[Y|X] = f(x)$. Some additional assumptions regarding the noise ϵ

are typically included in the model, such as independence and assumptions about its variance or distribution. Sometimes, for brevity, we will refer to a predictive model by its deterministic component $f : X \to \hat{Y}$, where $\hat{Y} = \mathbb{E}[Y|X]$.

Definition 6.2. *A* machine learning algorithm *is any computer algorithm that is able to automatically improve its performance of a task through experience.*

Supervised learning *refers to any machine learning algorithm that learns a predictive model* $f : X \to \hat{Y}$ *from a set of labeled data* $\mathcal{D} = \{(x_i, y_i)\}_{i=1}^{n}$. *The input X is often referred to as a* feature vector, *and its components as a set of* features. *The output Y is often referred to as the* target variable. *Each datum (x_i, y_i) is referred to as an* instance.

Unsupervised learning *is any machine learning algorithm that learns patterns in unlabeled data.*

Reinforcement learning *is any machine learning algorithm in which an agent takes a sequence of actions in an environment in order to maximize a cumulative reward.*

In Part II of this text, we will predominantly focus on predictive models. In Part III, we will explore machine learning algorithms more closely. As a quick illustration of the distinction laid out in Definition 6.2, consider the following.

Example 6.1. A computer algorithm is trained with a set of images as inputs, each input having a label "cat" or "dog." It is tested on its ability to recognize whether or not a new image is a cat or dog. This is an example of supervised learning, as the labels ("cat" and "dog") were provided with the training set.

A different algorithm is provided the same set of input images, except without being explicitly told what is a cat and what is a dog. It is asked to seek patterns in the data. Here, the algorithm might come up with patterns that are unexpected. For example, the algorithm might classify images of the side view versus images of a front view; or it might classify partially hidden versus fully visible animals. Or it might classify head shots versus body shots. Or it might classify based on whether or not the pet is wearing a collar. Or it might come up with its own concept, something that distinguishes the images that is present in the data but would not normally be recognized by the human eye. This is an example of unsupervised learning, as the algorithm is left to its own devices to infer patterns from the data.

Finally, our third algorithm has learned how to play the game *Go*. Here, the environment is the board. The actions is the set of permissible (legal) Go moves. And the reward is issued at the end of each game as +1 win or 0 for loss. The computer can be trained with actual Go games, and it will continue to learn based on its experience. That is, as it plays, it becomes better at playing. In fact, two computers can be set up to play each other, with entire games occurring at lightning speed, while the algorithm learns which strategies work best.   ▷

Predictive models themselves can be grouped into two different types, based on the type of target variable they are predicting.

Definition 6.3. *A* classification algorithm *is any supervised machine learning algorithm that learns a predictive model with a categorical output variable. The enumeration of possible outputs are referred to as* labels.

A regression algorithm *is any supervised machine learning algorithm that learns a predictive model with a continuous real output variable.*

In Example 6.1, the supervised algorithm is a classification algorithm, as it learned whether each picture represented a dog or a cat. An example of a regression algorithm is an algorithm that predicts home price based on a set of features (e.g., square footage, city, number of beds / baths, school district, etc.).

In general, the feature vector of a model will consist of a mix of continuous, categorical, and ordinal features. (An ordinal feature is similar to a categorical feature in the sense that it is finite and discrete, but it differs as it constitutes an ordered collection; e.g., *small, medium,* and *large*.) We will explore various requirements for data preprocessing for each of these feature types in Section 6.2.

Definition 6.4. *Given a predictive model $Y = f(X) + \epsilon$, with error term $\mathbb{E}[\epsilon] = 0$ and $\mathbb{V}(\epsilon) = \sigma^2$, an instance (x, y) is said to have a* weight *(or* prior weight*) $w > 0$ if the variance of its error ϵ is reduced by a factor w; i.e., if $y_i \sim f_Y$, then $\mathbb{E}[Y] = f(x_i)$ and $\mathbb{V}(Y - f(x_i)) = \sigma^2/w$.*

A weighted data set *$\mathcal{D} = \{(x_i, y_i, w_i)\}_{i=1}^n$ is a data set with a prescribed (known) set of prior weights, so that if $y_i \sim f_{Y_i}$, then $\mathbb{E}[Y_i] = f(x_i)$ and $\mathbb{V}(Y_i - f(x_i)) = \sigma^2/w_i$. The* total weight *$\omega$ is the sum of the individual weights; i.e., $\omega = \sum_{i=1}^n w_i$.*

An instance (x, y) with positive integer weight $w \in \mathbb{Z}^+$ is equivalent to a set of w identical instances $\{(x, y)\}_{i=1}^w$, so that weights are a convenient encapsulation for multiplicity.

We conclude this section with a more detailed definition on what constitutes a machine learning model.

Definition 6.5. *A* (supervised) machine learning model *over a feature space $\mathbb{F}$ into a target space $\mathbb{T}$ is a mapping*

$$f : \mathbb{D} \to \mathscr{F}(\mathbb{F}, \mathbb{T}), \tag{6.1}$$

where $\mathbb{D} = \mathscr{P}_{\text{fin}}(\mathbb{F} \times \mathbb{T})$ is the set of all possible finite subsets (or samples) of the data, and where $\mathscr{F}(\mathbb{F}, \mathbb{T})$ is the set of functions from $\mathbb{F}$ into $\mathbb{T}$.

Given a sample of data $\mathcal{D} \in \mathbb{D}$, known as training data, *the function $f_{\mathcal{D}} : \mathbb{F} \to \mathbb{T}$ (known as the* trained model*) is a predictive model for the target variable as it depends on the feature vector.*

If the target space is a finite set $\mathbb{T} = \mathcal{C}$ of categories, we say that the model is a classification *model. Otherwise, if the target space is a subset of the reals, $\mathbb{T} \subset \mathbb{R}$, we say that the model is a* regression *model.*

An unsupervised machine learning model *is similar, except that the set of data $\mathbb{D} = \mathscr{P}_{\mathrm{fin}}(\mathbb{F})$ consists only of the feature data.*

Another aspect of learning algorithms is that the mapping shown in Equation (6.1) need not be performed in a single step; i.e., the model can continuously *improve* as data are added. In this way, we can view a machine learning algorithm as a mapping

$$f : \mathbb{D} \times \mathscr{F}(\mathbb{F}, \mathbb{T}) \to \mathscr{F}(\mathbb{F}, \mathbb{T}),$$

which symbolizes the important aspect that the model can improve (or *learn*) with time.

6.1.2 Validation Metrics for Regression

In this section, we will consider a predictive model of the form $\mathbb{E}[Y|X] = f(X)$ that was trained using a regression algorithm. We will consider various metrics that can be used to measure the performance of the model.

Squared Errors

Consider a data set $\mathcal{D} = \{(x_i, y_i)\}$ and a model f with predictive values $\hat{y}_i = f(x_i)$. Note that the data set $\mathcal{D}$ is not necessarily the same data that was used to generate the model. We will discuss this in more detail in Section 6.1.5. We consider the following sum-of-squares statistics.

Definition 6.6. *Given a weighted data set $\mathcal{D} = \{(x_i, y_i, w_i)\}_{i=1}^{n}$ and model predictions $\hat{y}_i = f(x_i)$, we define the* sum of squared errors (SSE), *the* regression sum of squares (SSR), *and the* total sum of squares (SST) *as*

$$\mathrm{SSE} = \sum_{i=1}^{n} w_i (y_i - \hat{y}_i)^2, \tag{6.2}$$

$$\mathrm{SSR} = \sum_{i=1}^{n} w_i (\hat{y}_i - \overline{y})^2, \tag{6.3}$$

$$\mathrm{SST} = \sum_{i=1}^{n} w_i (y_i - \overline{y})^2, \tag{6.4}$$

respectively.

Note 6.1. The difference $y_i - \hat{y}_i$ between the true and predicted value is referred to as a *residual*. Therefore, many authors, especially in machine learning, refer to the sum of squared errors as the *residual sum of squares*, or RSS. We will use the phrases *sum of squared errors* and *residual sum of squares* interchangeably, however we favor the notation SSE as it is consistent with other sum-of-squares notation and the *mean-squared error* MSE, defined as MSE $=$ SSE$/n$ or, for a weighted data set, as MSE $=$ SSE$/\omega$.   ▷

One might be tempted to assume that the residual sum of squares and the regression sum of squares sum to the total sum of squares. This assumption, however, is not true in general. In fact, it is only true if the cross terms

$$\sum_{i=1}^{n}(y_i - \hat{y}_i)(\hat{y}_i - \overline{y})$$

sum to zero. Whether or not these terms vanish clearly depends on the model. For instance, consider the model that predicts $f(x) = 0$, for all x. Then the cross terms do not cancel, but instead sum to $-n\overline{y}^2$.

A common metric used in model evaluation is the coefficient of determination, which is often denoted simply and erroneously as R^2, which we define as follows.

Definition 6.7. *The* coefficient of determination (R^2) *is defined as*

$$R^2 = 1 - \frac{\text{SSE}}{\text{SST}}, \tag{6.5}$$

where SSE *and* SST *are defined as in Definition 6.6.*

This "R-squared" statistic really isn't the square of anything, other than its own square root, which need not be real. In other words, the R^2 statistic in certain cases can be negative. It cannot, however, be larger than 1. In fact, a value of $R^2 = 1$ would be a perfect model, as this is only achieved whenever $\hat{y}_i = y_i$, for all $i = 1, \ldots, n$. Hence we have the inequality

$$R^2 \leq 1.$$

In fact, the R^2 statistic *will* be negative whenever our predictions have a greater total error than had we simply predicted the average value $f(x) = \overline{y}$ for the entire data set. (See Exercise 6.1.) So a *positive* R^2 indicates we are doing better than no model at all (where, in absence of a model, we can simply use the average value as the value of all of our predictions).

The R^2 statistic is commonly interpreted as the *fraction of variance explained* by the model. To see this, note that the total sum of squares is essentially (a multiple of) the sample variance of the data. This represents the variance observed in the data set itself. The residual sum of squares, on the other hand, is how much variance is still *unexplained* by our model, as the residual sum of squares measures the variance between our predictions and the true values of the data.

The R^2 statistic is good in that the residual sum of squares is scaled by the total sample variance. Whereas interpretation of the MSE of a model completely relies on domain expertise, the R^2 statistic is a normalized metric that has a standard meaning.

A pitfall of the R^2 statistic, however, is that there is no universally adopted notion of which values of R^2 are acceptable. This is due, in part, to the fact that R^2 depends on the variance in the data. If we hold the mse fixed, the larger the variance in the data, the higher the value of R^2. In one context, an $R^2 = 0.20$ might be acceptable, but in another, an $R^2 = 0.90$ might not. For example, if we are modeling housing prices, which show a significant variability, a value of $R^2 = 0.90$ might translate to a RMSE in the tens of thousands of dollars, which would probably not be considered acceptable to the given application (example from Kuhn and Johnson [2013]).

The R^2 statistic is, however, useful in comparing different models. But even then there is an important caveat: the R^2 statistic will increase as one adds explanatory variables in the model. This leads to the erroneous conclusion that one should keep adding features to the model: everything except the kitchen sink. One important modification that addresses this is the following.

Definition 6.8. *The* adjusted coefficient of determination $\bar{R}^2$ *is defined as*

$$\bar{R}^2 = 1 - \frac{\text{SSE}/\text{df}_e}{\text{SST}/\text{df}_t}, \tag{6.6}$$

where $df_t = n - 1$ *is the number of degrees of freedom of the estimate of the population variance and* $df_e = n - p$ *is the number of degrees of freedom of the estimate of the residual variance, where* p *is the total number of independent modeling parameters of the model.*

Unlike regular R^2, the adjusted $\bar{R}^2$ will increase with the inclusion of additional features only if the the model improvement is more than one would expect to see by chance.

An alternate assessment metric is given as follows.

Definition 6.9. *The* mean absolute error (MAE) *of a model is given by*

$$MAE = \frac{1}{n} \sum_{i=1}^{n} |\hat{y}_i - y_i|. \tag{6.7}$$

Similarly, the mean absolute percentage error (MAPE) *of a model is given by*

$$MAPE = \frac{1}{n} \sum_{i=1}^{n} \left| \frac{\hat{y}_i - y_i}{y_i} \right|. \tag{6.8}$$

Essentially, MAPE is a measure of how close the predictions are to the true values. A MAPE of 10% indicates that the predictions are typically about $\pm 10\%$ of the true values. This metric also has various flaws: it cannot be used if there are any zeros in the data, and there is an asymmetry between under-predictions (which cannot exceed 100%) and over-predictions (which are unlimited).

Bias–Variance Tradeoff

We will discuss model selection in Section 6.1.5. Presently, however, we shall content ourselves with laying out some of the theoretical groundwork. We begin with two definitions.

Definition 6.10. *Given a normal predictive model $Y = f(X) + \epsilon$, for $\epsilon \sim N(0, \sigma^2)$, where $f(X)$ is an unknown function of the input X. Let $\hat{f}(X) = \hat{f}(X; \mathcal{D})$ be an estimate for the unknown function $f(X)$ that is learned on a set of data $\mathcal{D} = \{(X_i, Y_i)\}_{i=1}^{n}$. Then we define the* bias *of the model $\hat{f}$ at a point $X = x$ as*

$$\text{bias}\left[\hat{f}(x)\right] = \mathbb{E}\left[\hat{f}(x; \mathcal{D})\right] - f(x). \tag{6.9}$$

Similarly, we define the model variance at a point $X = x$ as

$$\mathbb{V}\left[\hat{f}(x)\right] = \mathbb{E}\left[\left(\hat{f}(x; \mathcal{D}) - \mathbb{E}[\hat{f}(x; \mathcal{D})]\right)^2\right]. \tag{6.10}$$

In Equations (6.9) and (6.10), all expectations are taken relative to the data $\mathcal{D} \sim f_{X,Y}^n$.

Definitions in hand, we are now ready to state our main result.

Theorem 6.1 (Bias–Variance tradeoff). *Let $Y = f(X) + \epsilon$, where $\epsilon \sim N(0, \sigma^2)$, and $\hat{f}(X) = \hat{f}(X; \mathcal{D})$ be as in Definition 6.10. Then the expected squared residual at a point $X = x$ is given by*

$$\mathbb{E}\left[\left(Y - \hat{f}(x; \mathcal{D})\right)^2\right] = \sigma^2 + \text{bias}\left[\hat{f}(x)\right]^2 + \mathbb{V}\left[\hat{f}(x)\right], \tag{6.11}$$

where the expectation on the left-hand side is taken over the joint distribution for Y and $\mathcal{D}$.

Note 6.2. The term σ^2 is often referred to as the *irreducible error*, as it is due to the white noise $\epsilon \sim N(0, \sigma^2)$ in the random variable Y; i.e., it is inherent uncertainty that cannot be remedied by modeling.   ▷

Proof. We begin by expanding the residual for fixed $X = x$ as

$$Y - \hat{f}(x; \mathcal{D}) = (Y - f(x)) + \left(f(x) - \mathbb{E}\left[\hat{f}(x; \mathcal{D})\right] \right) + \left(\mathbb{E}\left[\hat{f}(x; \mathcal{D})\right] - \hat{f}(x; \mathcal{D}) \right).$$

Now, the first term depends on the random variable Y, whereas the second and third terms depend on the random variable $\mathcal{D}$. Therefore, the first term is independent with the second and third terms, and we conclude that

$$\mathbb{E}\left[(Y - f(x)) \left(f(x) - \mathbb{E}\left[\hat{f}(x; \mathcal{D})\right] \right) \right] = 0$$

and

$$\mathbb{E}\left[(Y - f(x)) \left(\mathbb{E}\left[\hat{f}(x; \mathcal{D})\right] - \hat{f}(x; \mathcal{D}) \right) \right] = 0.$$

The second term is actually independent of $\mathcal{D}$ itself, so a similar result holds for the penultimate and final terms

$$\mathbb{E}\left[\left(f(x) - \mathbb{E}\left[\hat{f}(x; \mathcal{D})\right] \right) \left(\mathbb{E}\left[\hat{f}(x; \mathcal{D})\right] - \hat{f}(x; \mathcal{D}) \right) \right] = 0.$$

By squaring and taking expectation over our expression for the residual, and by recalling that $Y - f(x) = \epsilon$, we therefore arrive at the expression

$$\mathbb{E}\left[\left(Y - \hat{f}(x; \mathcal{D}) \right)^2 \right] = \mathbb{E}[\epsilon^2] + \left(\mathbb{E}\left[\hat{f}(x; \mathcal{D})\right] - f(x) \right)^2$$
$$+ \mathbb{E}\left[\left(\hat{f}(x; \mathcal{D}) - \mathbb{E}\left[\hat{f}(x; \mathcal{D})\right] \right)^2 \right].$$

The result follows. □

Note 6.3. The *mean-squared error* MSE of a model can be defined by taking the expectation of Equation (6.11) over the random variable X.   ▷

We will see illustrations of this result as we progress through the chapter. Note that the bias-variance theorem is inherently concerned with regression problems, as it is a statement concerning the our expectation (literally) for the residual sum of squares.

6.1.3 Validation Metrics for Classification

In classification problems, the target variable Y constitutes a categorical response, taking one of c values in a set $\mathcal{C}$, whose elements can be labeled as $1, \ldots, c$. Oftentimes we will directly model a set of functions $f_i(X)$, for $i = 1, \ldots, c$, that are required to satisfy the constraints that $0 < f_i(X) < 1$, for all $i = 1, \ldots, c$ and $X \in \mathcal{X}$, and the normalization condition

$$\sum_{i=1}^{c} f_i(X) = 1.$$

Moreover, these functions should satisfy the property that the larger the value of $f_i(X)$, for a given class label i, the more likely it is that $Y = i$. They can be thought of as probabilities, but these outputs typically do not correspond to the actual real probability. When the raw functions $f_i'(X)$ do not satisfy the normalization conditions, we can always apply the *softmax transformation*

$$f_i(X) = \frac{e^{f_i'(X)}}{\sum_{i=1}^{c} e^{f_i'(X)}}. \qquad (6.12)$$

(See Exercise 6.2.) However, our goal is not to predict a number between zero and one, but to predict class labels for a new instance X. A common method is to simply select the class with the largest value of f_i:

$$\hat{Y}(X; \mathcal{D}) = \arg\max_{i=1,\ldots,c} f_i(X; \mathcal{D}). \qquad (6.13)$$

However, in many cases, we will be specifically considering classification algorithms for the purpose of predicting a *binary response*, so that $Y \in \{0, 1\}$. In this case, we only need to specify a single function $f(X) = f_1(X)$. In this case, it is more common to use a threshold t to determine the final classification:

$$\hat{Y}_t(X; \mathcal{D}) = \mathbb{I}[f(X; \mathcal{D}) > t]. \qquad (6.14)$$

Here, the same model will yield a different set of predictions for different thresholds. We will discuss this in more depth at the end of this subsection.

Note 6.4. We should further note that many classification algorithms *do not model an actual probability*! That is, the function $f(X)$ that is learned by the classification algorithm will satisfy the properties of a probability, but will not correspond to a probability itself. A value of $f(X) = 0.5$ means that the response $Y = 1$ is more likely than it would have been if $f(X) = 0.4$, but it does *not* mean that there is a 50% *probability* that the response is $Y = 1$. We will discuss this concern, and its remedy, in Section 6.1.4.   ▷

Standard Classification Metrics

For the remainder of this section, we shall consider the common case of binary classification problems. In this context, the response $Y = 1$ is referred to as "the event," and it typically corresponds to the target of our prediction task. Common applications include a diagnostic for a lab test and a user in a freemium mobile app making a purchase. This distinction is made without loss of any generality, as $Y = 0$ and $Y = 1$ could just as easily correspond to dogs and cats. However, for the purpose of the following, we will consider the event $Y = 1$ to be the target of our prediction problem.

Definition 6.11. *Given a classification problem and a set of true values Y_i and predictions $\hat{Y}_i$, for $i = 1, \ldots, n$, where each Y_i may take one of m distinct values, the confusion matrix C is the $c \times c$ matrix of counts, where*

$$C_{ij} = \sum_{k=1}^{n} \mathbb{I}\left[\hat{Y}_k = i \text{ and } Y_k = j\right].$$

Thus, in a confusion matrix, the rows correspond to the predicted classes, and the columns correspond to the actual (true) classes. For the case of binary classification, it is conventional (for some reason) to write the rows and columns of the confusion matrix in *descending* order[1]. A typ-

	$Y = 1$	$Y = 0$
$\hat{Y} = 1$	TP	FP
$\hat{Y} = 0$	FN	TN

Table 6.1: Confusion matrix for binary classification.

ical confusion matrix is shown in Table 6.1. The elements of the matrix are labeled TP, FP, FN, and TN. These correspond to true/false positive/negative, as defined below.

Definition 6.12. *In a binary classification problem, the elements of the confusion matrix are referred to as*

1. True Positive (TP)—*the number of correct positive predictions ($\hat{Y} = 1$ and $Y = 1$),*
2. False Positive (FP)—*the number of incorrect positive predictions ($\hat{Y} = 1$ and $Y = 0$),*
3. False Negative (FN)—*the number of incorrect negative predictions ($\hat{Y} = 0$ and $Y = 1$), and*
4. True Negative (TN)—*the number of correct negative predictions ($\hat{Y} = 0$ and $Y = 0$).*

Note 6.5. In Definition 6.12, think of the terminology in the context of a laboratory test for a medical condition: a positive result means you tested positive for the condition and a negative result means you tested negative for the condition. The True/False indicates whether or not the test result (the predicted value) was correct.   ▷

The first validation metric for a classifier is accuracy.

Definition 6.13. *The* accuracy *of a classifier is the ratio of the number of correct predictions to the number of overall predictions, i.e.,*

$$accuracy = \frac{TP + TN}{TP + FP + FN + TN}.$$

However, considering accuracy alone can be misleading.

---
[1] I suppose otherwise there wouldn't be anything confusing about it.

Example 6.2. Consider a binary classification problem for a rare event. Now consider the classifier that classifies *everything* as $\hat{Y} = 0$. The confusion matrix for 100 samples is given in Table 6.2. The accuracy of this model on

	$Y = 1$	$Y = 0$
$\hat{Y} = 1$	0	0
$\hat{Y} = 0$	1	99

Table 6.2: Confusion matrix for Example 6.2; 99% accuracy.

this data set is 99%, even though the model (by design) doesn't do anything at all.   ▷

To remedy this, statisticians typically refer to two quantities known as sensitivity and specificity, defined below.

Definition 6.14. *The* sensitivity *and* specificity *of a binary classifier are the ratios of the number of true positives or negatives, respectively, to the total number of positives or negatives, respectively; i.e.,*

$$sensitivity = \frac{true\ positives}{total\ actual\ positives} = \frac{TP}{TP + FN} = TPR,$$

$$specificity = \frac{true\ negatives}{total\ actual\ negatives} = \frac{TN}{FP + TN} = TNR = 1 - FPR.$$

The sensitivity is equivalent to the true positive rate *(TPR) and the specificity is equivalent to the* true negative rate *(TNR), which is equivalent to one minus the* false positive rate *(FPR).*

It is desirable to have high sensitivity and specificity, and a low false positive rate. However, these two metrics still do not capture the full picture, as illustrated in our next example.

Example 6.3. Consider a binary classification problem with confusion matrix given in Table 6.3. The overall accuracy ($190/210 \approx 90.47\%$) is good.

	$Y = 1$	$Y = 0$
$\hat{Y} = 1$	10	20
$\hat{Y} = 0$	0	180

Table 6.3: Confusion matrix for Example 6.3; 100% sensitivity, 90% specificity.

Similarly, the sensitivity and specificity both look healthy:

$$\text{sensitivity} = \frac{10}{10} = 100\%,$$

$$\text{specificity} = \frac{180}{200} = 90\%.$$

However, what does a positive test result $\hat{Y} = 1$ mean? Given a positive test result, we have to look at the first row of the confusion matrix. And we see that we only have a 33% probability of actually being a true positive. This discrepancy is not captured at all when looking at sensitivity and specificity alone.

$\triangleright$

As a result of Example 6.3, in the field of machine learning, it is more common to analyze the following pair of metrics.

Definition 6.15. *The* precision *and* recall *of a binary classifier are defined as*

$$precision = \frac{true\ positives}{total\ predicted\ positives} = \frac{TP}{TP + FP},$$

$$recall = \frac{true\ positives}{total\ actual\ positives} = \frac{TP}{TP + FN}.$$

Recall is the same as sensitivity.

In Example 6.3, the precision is 33% and the recall is 100%. Thus, a positive example has a 100% chance of being identified as positive; however, an example with a positive prediction only has a 33% chance of being an actual positive.

If a combined metric is required, one can use the F_1-score, which is the harmonic mean of precision and recall. More generally, we can use F_β, for $\beta > 0$, which is defined as

$$F_\beta = (1 + \beta^2)\frac{\text{precision} \cdot \text{recall}}{\beta^2 \text{precision} + \text{recall}}.$$

For the case $\beta = 1$, we recover the harmonic mean of the two metrics.

Receiver-operator Characteristic (ROC) Curves

As stated above, the decision of whether or not to classify the output $f(X)$ of a binary classifier as an event is typically determined by setting a threshold $t \in [0,1]$ and applying Equation (6.14). All of our validation metrics thus far apply to a single predictive model, i.e., following the selection of an appropriate threshold. But how should we compare different classifiers without regard to the particular value of the threshold parameter? The answer lies in the following.

Definition 6.16. *The* receiver–operator characteristic (ROC) *curve of a predictive model* $f(X; \mathcal{D})$, *with classification function*

$$\hat{Y}_t(X; \mathcal{D}) = \mathbb{I}\left[f(X; \mathcal{D}) > t\right],$$

for $t \in [0, 1]$, *is the parametric curve* $\mathbf{r} : [0, 1] \to [0, 1]^2$ *defined by*

$$\mathbf{r}(t) = \langle FPR(t), TPR(t) \rangle,$$

where $FPR(t)$ *and* $TPR(t)$ *are the false- and true-positive rates as a function of the threshold parameter* t.

Typically, the x-label of the ROC curve is represented as "1 - specificity," which, of course, is equivalent to the false-positive rate. All ROC curves satisfy the endpoint conditions. For an example ROC curve, see Figure 7.3.

Proposition 6.1. *Let* $\mathbf{r} : [0, 1] \to [0, 1]^2$ *be an ROC curve. Then*

$$\mathbf{r}(0) = \langle 1, 1 \rangle \qquad and \qquad \mathbf{r}(1) = \langle 0, 0 \rangle.$$

Moreover, the TPR is a nondecreasing function of the FPR.

Proof. We begin by considering the threshold $t = 0$. In this case, $\hat{Y}_0(X) = 1$, for all X, since the function $f \in (0, 1)$. Therefore

$$TPR(0) = FPR(0) = 1.$$

Similarly, at the endpoint $t = 1$, we have $\hat{Y}_1(X) = 0$, for all X, so that

$$TPR(1) = FPR(1) = 0.$$

This proves the two endpoint conditions; i.e., the ROC curve connects the endpoints $\langle 0, 0 \rangle$ and $\langle 1, 1 \rangle$.

Next, let us examine the monotonicity of the true-positive rate, when cast as a function of the false-positive rate. Consider any two t_1, t_2, with $0 \le t_1 < t_2 \le 1$. By increasing the threshold from t_1 to t_2, any instance X with $f(X; \mathcal{D}) \in (t_1, t_2)$ will no longer be classified as a positive example. The total of each column of the confusion matrix must remain constant; however, counts will drip from row $i = 1$ into row $i = 0$. Thus, we have

$$TPR(t_2; x) \le TPR(t_1; x) \qquad and \qquad FPR(t_2; x) \le FPR(t_1; x).$$

The result follows. □

There are two interesting points to note. The first is that the line $TPR = FPR$ represents a model that classifies at random. That is, if the ranking of the model outputs $f(X_1) < f(X_2) < \cdots < f(X_n)$ is random, then the probability of classifying an instance as a positive is independent of whether or not the instance actually is positive. Thus, we should expect that the actual positive instances are classified as positive (TPR) at the same rate as are the actual negative instances (FPR). This gives some guidance on interpreting the ROC curve. Curves that fall below the diagonal "$y = x$" line are fairing worse than had you just assigned probabilities at random. Curves at are entirely above the diagonal are performing better than average.

The second point is that a perfect model would have $TPR = 1$ and $FPR = 0$. Thus, the closer the curve gets to the point $\langle 1, 0 \rangle$, typically the better the model. Stated differently: the area under the curve (which is bound between 0 and 1) is a measure of predictive performance for the model. We therefore defining the following.

Definition 6.17. *The* area under the curve (AUC) *of a classifier is defined as the area bounded by the ROC curve and the lines $TPR = 0$ and $FPR = 1$.*

Clearly, it follows that $0 < AUC < 1$, for any model. Moreover, an $AUC = 1/2$ is a model that performs as good as random. And a model with an AUC close to 1 is a superior model.

6.1.4 Predicting Class Probabilites

As mentioned in Note 6.4, the function $f(X; \mathcal{D})$, produced by a binary classifier, is not a true probability, though it possesses the basic properties of probabilities. We now turn to the case where the goal of our classifier is actually to determine an accurate *probability*, as opposed to an actual label.

Calibration

A classification model is said to be *well calibrated* if its output values $f(X)$ represent the probability that $Y = 1$; i.e., a value $f(X) = 0.3$ actually means that the given instance has a 30% probability of being positive. In order to visualize how well calibrated a model is, we rely on the following.

Definition 6.18. *Given a set of data $\{X_i, Y_i\}_{i=1}^{n}$ and prediction values $f(X_i)$ of a classification algorithm. The instances are ordered relative to their prediction values, so that $f(X_1) < f(X_2) < \cdots < f(X_n)$. Then they are divided into k buckets or bins, of approximately equal size n/k. Next, let f_i represent the average prediction value of the instances in the ith bucket, and let p_i represent the fraction of positive instances within the ith bucket. Then the curve connecting the points $\langle f_i, p_i \rangle$ constitutes a* calibration curve *for the model.*

Note 6.6. The diagonal "$y = x$" line represents a perfectly calibrated model, as the prediction values are equivalent to the true probabilities.   ▷

Note 6.7. An alternative formulation of the calibration curve plots the bin number on the x-axis and concurrently plots two separate curves on the y-axis: one for the predicted values and one for the true values of each bin. This format is suitable for viewing the "calibration" of a regression model as well, and can serve as a visual diagnostic as to whether a model is well calibrated. By comparing the true and predicted curves concurrently, it is also more interpretable to business stakeholders.   ▷

Typically, a calibration curve will have a sigmoidal shape. We can therefore adjust the prediction values by fitting the calibration curve to the sigmoid function

$$p_i = \frac{1}{1 + \exp(-\beta_0 - \beta_1 f_i)}.$$

Applying the same curve to each instances prediction value $f(X)$ therefore returns a prediction value that is better calibrated as a probability. This method was introduced by Platt [2000]. This approach and an alternative approach, isotonic regression, are discussed in Niculescu-Mizil and Caruana [2005]. Isotonic regression is especially suitable when the calibration curve does not have a sigmoidal shape.

Brier Score and Cohort Probabilities

We next discuss a metric used to assess probability models, due to Brier [1950].

Definition 6.19. *The* Brier score *for a binary classifier is defined by*

$$\text{BS} = \frac{1}{n} \sum_{i=1}^{n} (y_i - p(X_i; \mathcal{D}))^2, \tag{6.15}$$

where y_i is the actual value and $p(X_i; \mathcal{D})$ is the predicted probability for of the ith instance.

The Brier score differs from the residual sum of squares, as the residual sum of squares would be defined based on the final classification—and not the predicted probability—of the each instance. The prediction probabilities are typically generated from the prediction values $f(X_i; \mathcal{D})$ of a classifier by implementing either Platt scaling or isotonic regression, as discussed in the preceding paragraph.

The Brier score has an interesting decomposition when the prediction probabilities are made for fixed *cohorts*, or groups with similar characteristics, as opposed to at the individual level. This is common when there is a finite and manageable number of permutations of the feature set.

Proposition 6.2. *Suppose a unique probability is provided for each of k cohorts. Then the Brier score is equivalent to the following two-component decomposition*

$$\text{BS} = \frac{1}{n} \sum_{i=1}^{k} n_i \left(p_i - \bar{y}_i \right)^2 + \frac{1}{n} \sum_{i=1}^{k} n_i \bar{y}_i \left(1 - \bar{y}_i \right), \qquad (6.16)$$

where n_i is the count of instances in the ith cohort, and $\bar{y}_i$ is the observed event probability in the ith cohort.

The first term in Equation (6.16) is related to the calibration: how well the cohort predicted probabilities align with the observed true values. The second term is an expression of the average inherent uncertainty within each cohort. This second term represents an irreducible error, as it cannot be changed with model improvements, for a fixed set of cohorts. We leave the proof to the reader. (See Exercise 6.3.)

6.1.5 Validation Methodology

Over the preceding pages, we were a bit nonchalant regarding to the data to which each of the formulas should be applied. We now seek to remedy that coolness by discussing this issue in greater depth. For additional references, see Hastie, *et al.* [2009] and Kuhn and Johnson [2013].

Loss Functions

We begin with a discussion on *loss functions*. This will allow us to speak generally in our conversation on error, without regard to the particular type of predictive model we are addressing.

Definition 6.20. *In a predictive learning model, any function of the form* $L : \mathbb{R}^2 \to [0, \infty)$ *is called a* loss function *if it has the properties that*

$$L(y, y) = 0$$

and $L(Y, \hat{Y}(X)) \to 0$ *as* $\hat{Y}(X) \to Y$, *for any instance* $(X, Y) \sim f_{X,Y}$.

For regression problems, common loss functions include squared error and absolute error,

$$L(y, \hat{y}) = (y - \hat{y})^2 \qquad (6.17)$$
$$L(y, \hat{y}) = |y - \hat{y}|, \qquad (6.18)$$

respectively.

For a c-class classification problems, where our random target variable $Y \in \mathcal{C}$, *common loss functions include binary loss and log-loss,*

$$L(y, \hat{y}) = \mathbb{I}[y \neq \hat{y}] \qquad (6.19)$$

$$L(y, p) = -\sum_{k=1}^{c} \mathbb{I}[y = k] \log p_k = -\log p_y, \qquad (6.20)$$

respectively.

Unlike Equations (6.17)–(6.19), Equation (6.20) is a function of the predicted probability of a classification model, not the class predictions itself. Here, p_k is the probability that a given instance belongs to class k, for $k = 1, \ldots, c$. Suppose that the instance does indeed belong to class k, for some particular $k \in \{1, \ldots, m\}$. Then the larger the value of p_k, the better the performance of the model. Recall from basic logarithm properties, that $-\log p_k > 0$ and $-\log p_k \to 0$ as $p_k \to 1$. The smaller the predicted value p_k, the worse the model did (since $y = k$ is correct), and the larger the value of $-\log p_k$.

The loss function Equation (6.17) corresponds to the residual sum of squares from Equation (6.2), whereas the loss function Equation (6.18) corresponds to the MAE from Equation (6.7). Similarly, the loss function Equation (6.19) is related to accuracy; in fact, it is 1 minus the accuracy.

Finally, we note that, for binary classification problems, if we let p represent the predicted probability of the positive label (typically, the minority label), then the log loss of Equation (6.20) simplifies as

$$L(y, p) = -y \log p - (1 - y) \log(1 - p). \tag{6.21}$$

Training, Test, and Prediction Errors

When applying any of the above metrics (e.g., Equations (6.2) and (6.15)) to the training set $\mathcal{D}$ itself, the result is referred to as the *training error*

$$\text{ERR}_\mathcal{D} = \sum_{i=1}^n L(y_i, f(x_i, \mathcal{D})). \tag{6.22}$$

This, however, is not a good metric with which to assess a model, because the model was able to train on the results that we are testing its performance on. This can lead to *overfitting* of a model, which occurs when a model has a low training error but generalizes poorly to new data. To remedy this, we next define several types of errors, that are determined based on how they are applied, but not the individual context to which they are applied. Thus, we will develop the following for a generic loss function, which will depend on the particular context.

Definition 6.21. *In a predictive learning model $f(X; \mathcal{D})$, learned from a data set $\mathcal{D}$, with loss function L, the* test error *or* generalization error *is the expected error over an independent sample*

$$\text{ERR}_\mathcal{D} = \mathbb{E}\left[L(Y, f(X; \mathcal{D}))\right]. \tag{6.23}$$

Here, the expectation is with respect to the random variable $(X, Y) \sim f_{X,Y}$.

The test error is thus the expected loss on an *independent sample*. Note that the test error is for a particular trained model $f(\cdot; \mathcal{D})$. The predictive model, and thus the test error, might have come out differently had the algorithm learned from an alternate training set $\mathcal{D}$. Thus, measuring test error doesn't go far enough to properly assess our model: what if our model had been learned from a different training set?

Definition 6.22. *In a predictive learning model* $f(X; \mathcal{D})$, *which can be learned from any data set* $\mathcal{D}$, *with loss function* L, *the* prediction error *is the* expected test error

$$\text{ERR} = \mathbb{E}\left[\text{ERR}_{\mathcal{D}}\right], \tag{6.24}$$

where, the expectation is with respect to the training set $\mathcal{D}$.

Thus, the test error is the expected loss given our particular training set, and the prediction error is the expected test error of a model, without regard to the particular training set deployed. Clear as pudding[2].

Model Selection and Assessment

A predictive model does not occur in a vacuum. Commonly, we are never interested in a single model, but a family of related models that follow a similar methodology. That is to say, predictive models are typically dependent on a number of *tuning parameters* or *hyperparameters* that specify the model or how it operates. The tuning parameters may vary the complexity of the model, or otherwise determine which features should be used to train the model. Once we have determined (what we believe to be) the optimal model, we then want to determine an accurate measure of its expected performance.

We therefore find ourselves faced with two distinct tasks:

- *Model selection*: estimate the performance of the various models under consideration in order to select the best one;
- *Model assessment*: evaluate the performance of the selected model.

At first pass, the task of model assessment might seem redundant; after all, have we not already estimated each model's performance before making our selection? Such an approach, however, often leads to folly, as it fails to account for any *selection bias* incurred during the model selection phase.

Example 6.4. A set of ten models are trained on a training set $\mathcal{D}$, yielding $f_i(X)$ for $i = 1, \ldots, 10$. The models are then applied to a separate test set $\mathcal{T}$, for which the test error is estimated using the mean-squared error ($\text{MSE} = \text{SSE}/df$).

Now, suppose that each model is exactly equivalent, except that they may perform differently on different data sets. That is, suppose that the

---

[2] Perhaps I'll start a new trend of ending mathematical statements with CAP, like proofs are ended with QED.

prediction error for each of our ten models is exactly $\mathrm{ERR} = 10$, and the model variances are all $\mathbb{V}(f_i(x)) = 9$, for each $i = 1,\ldots,10$. We can simulate such a scenario by drawing a random sample of ten numbers from $N(10,9)$, obtaining the values

$$8.72,\ 9.44,\ 10.13,\ 6.75,\ 10.59,\ 11.67,\ 8.61,\ 6.35,\ 11.81,\ 2.76.$$

Upon seeing these results, we have a clear winner: model 10. (Even though, in reality, the models are exactly identical.)

Now, if we stop there, and report that our model has an expected generalization error of 2.76, we have a problem. This is called selection bias, and it illustrates the importance of the model assessment task as a separate task from the model selection process.   ▷

To address the issue of selection bias in model selection and assessment, we commonly divide our data set into three subsets:

1. A *training set* $\mathcal{D}$: the data set used to train the models;
2. A *validation set* $\mathcal{V}$: the data set used to estimate the test error of the various models, used for model selection;
3. A *test set* $\mathcal{T}$: the data set used to estimate the generalization error for the final model.

A typical split may be 50–25–25, but this in reality depends on the application and the size of data. Several tools that can be used for data sets that are two small to be amenable to such a split are discussed in Hastie, *et al.* [2009].

Cross Validation

Another approach often used to estimate the prediction error of a model is k-fold cross validation. The idea is that we divide our data set randomly into k partitions (or folds). We then proceed to train our model k times, each time we leave out one of the folds as a test set. This method therefore yields k separate estimates of the test error, which can be averaged together to estimate the expected test error.

Let $\mathcal{D} = \{(X_i, Y_i)\}_{i=1}^n$ be our data set, and let $\kappa : \{1,\ldots,n\} \to \{1,\ldots,k\}$ be a function that randomly assigns each data point into one of k bins (or folds), and let $\kappa^{-1}(j) = \{i : \kappa(i) = j\}$. Now define

$$\mathcal{D}_j = \{(X_i, Y_i)\}_{\kappa(i) \neq j} \qquad \text{and} \qquad \mathcal{T}_j = \{(X_i, Y_i)\}_{\kappa(i)=j}$$

be the set with the jth fold removed. We want to train our model on $\mathcal{D}_j$ and test our model on $\mathcal{T}_j$, for $j = 1,\ldots,k$. We may estimate the test error for the model trained on the set $\mathcal{D}_j$ as

$$\hat{\mathrm{ERR}}_{\mathcal{D}_j} = \frac{1}{|\kappa^{-1}(j)|} \sum_{i \in \kappa^{-1}(j)} L(Y_i, f(X_i; \mathcal{D}_j)), \tag{6.25}$$

for $j = 1, \ldots, k$. By averaging these results, we thus arrive at an estimate for the prediction error

$$\hat{\text{ERR}} = \frac{1}{k} \sum_{j=1}^{k} \hat{\text{ERR}}_{\mathcal{D}_j}. \tag{6.26}$$

Similarly, we can compute the sample variance of our k estimated test errors to estimate the model's variance. An illustration of five-fold cross validation is shown in Figure 6.1.

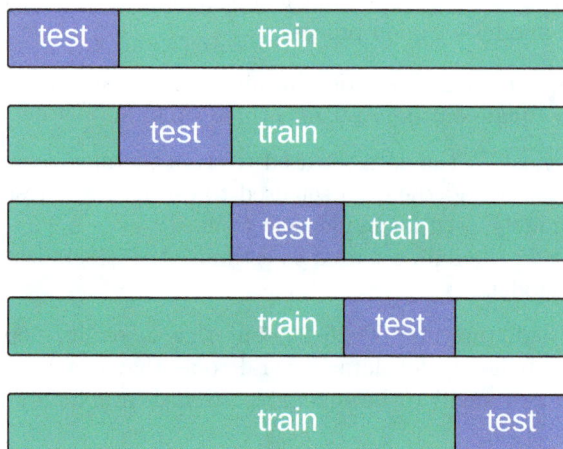

Fig. 6.1: An illustration of five-fold cross validation.

A typical choice of k may be $k = 5$ or $k = 10$. The case $k = n$ is referred to as *leave-one-out cross validation* or the *jackknife*.

6.1.6 Validation on Temporal Datasets

In Section 5.4, we discussed various kinds of stochastic processes that generate data over time. In practice, we must take extra care to perform validation on data that has an explicit time component. There are two potential problems that can arise. First, even when traditional machine-learning validation methodologies are available (e.g., cross validation), which requires that we have a well defined target variable in our training data, user behavior can still vary over time, so that a model that is valid today might not have been valid yesterday. Second, we might encounter data that is not *fully baked*, so that there is no well defined target variable. This occurs with survival processes (Section 5.3), as well as in customer lifetime value models (Section 10.3). In these models, we are learning from behavioral data we have *to date*, in order to predict future behavior. In both cases, it is therefore important to perform a historical validation over a stretch of time.

Before jumping into how we are going to validate such a model, let us first take a moment to examine how our model will be deployed in production. In order to run our model in production, we will require a well defined *training window* and a *prediction range*. In addition, we often desire to buffer the immediately trailing data from our training window. Formally, we define these as follows.

Definition 6.23. *When deploying a model over a* temporal dataset, *or a dataset with an explicit time component, the* training window *is the date range for the training set, and the* prediction range *is the date range for the prediction set.*

These are commonly specified relative to a point in time, *which defaults to the present date, and three configuration parameters*

1. delay: *the number of days between the end of the training set and the point in time;*
2. window: *the length (in days) of the training set;*
3. range: *the length (in days) of the prediction set.*

Given these definitions, our training window and prediction range can be expressed as

$$\text{training window} = [\texttt{point\_in\_time} - \texttt{delay} - \texttt{window}, \texttt{point\_in\_time} - \texttt{delay})$$
$$\text{prediction range} = [\texttt{point\_in\_time} - \texttt{range}, \texttt{point\_in\_time}).$$

This is shown schematically in Figure 6.2. Here, the training set consists of the green data, whereas the prediction set is the blue data. In the figure, the delay and range are shown as equal, though this is not a requirement. In production, the "point in time" is always taken to be the present date.

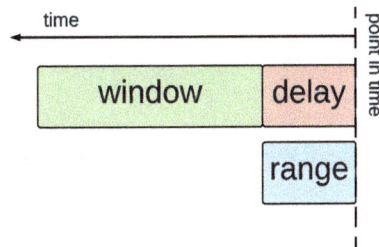

Fig. 6.2: Training (green) and prediction (blue) data for temporal datasets. In production, *point in time* is the current date.

So why don't we just say "now" instead of "point in time"? The beauty of this design, is that we are treating "now" as a parameter. This means that we can pass a "different now" into our runtime job in order to see what our model results would have been had we run our model, with the same configuration, at some point in time in the past.

We can take advantage of this configuration to perform a *backtesting validation* (or, simply, *backtest*), which consists of running our model using historical data at various points in time in the past, in order to write predictions over a period of time. Not only can we run our standard validation metrics to determine how well our model would have performed, but we can also see how our performance varied over our stretch of time. For example, we can determine how much our model predictions might fluctuate in time. A backtesting validation with four jobs is shown in Figure 6.3.

Note that each job is staggered by the range, so that we have full coverage of our predictions over a substantial period of time. In other words,

$$\texttt{point\_in\_time(job\_i)} = \texttt{point\_in\_time} - \texttt{range} \times i,$$

where $i \in 0, \dots, n-1$, where n is the number of backtest jobs deployed, and $\texttt{point\_in\_time}$ is the ending point in time.

Backtesting is especially critical for online process data, which is commonly not fully baked at the time we write our predictions (e.g., survival processes). For such a scenario, we can set the point in time for $\texttt{job-0}$ as the latest point in time for which fully baked data is available. The model still trains using only the data it would have had available at that point in time, but we can use the fully baked data from those cohorts in order to validate our predictions. In such a scenario, cross validation is not necessary. We are training our model over a historical backtesting period in the exact same way that we would train the model during production. We can therefore see exactly how well our model would have fared over our historical backtesting period, had it been live and in production at the time.

6.2 Preprocessing

It is seldom the case that one can use the features (independent variables) to train a model in their raw form. Rather, transformations are commonly applied prior to feeding the features into the model. Now, what is good for the goose, is good for the gander: transformations applied to the training set should equally be applied to the validation and test sets prior to computing the model's predictions.

Most of this section will rely heavily on the scikit-learn package. For an introduction to scikit-learn in machine learning, see Pedregosa, *et al.* [2011]. For a discussion of the design principles of its API, see Buitinck, *et al.* [2013]. And, of course, for the latest and up-to-date changes on functionality, visit the scikit-learn user guide at

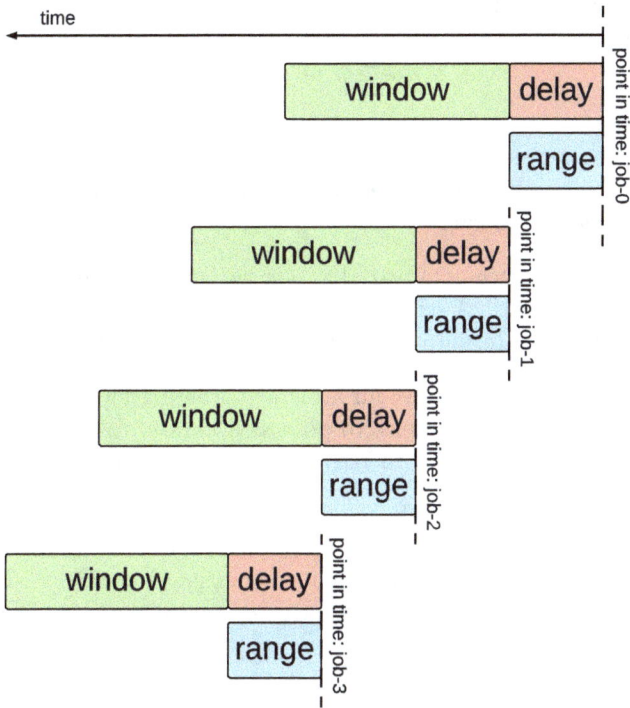

Fig. 6.3: Backtesting validation for temporal datasets.

https://scikit-learn.org/stable/user_guide.html.

6.2.1 Categorical Features

A *categorical variable* is a variable that can take on any value from a finite set $\mathcal{C}$. We will consider two types of categorical features: ordinal and nominal. To differentiate between these two, we introduce the following.

Definition 6.24. *Let $\mathcal{S}$ be a set and "$\leq$" a binary relation over $\mathcal{S}$[3]. Then the binary relation is a* total order *if it satisfies*

1. Connexity: $x \leq y$ *or* $y \leq x$, *for all* $x, y \in \mathcal{S}$;
2. Reflexivity: $x \leq x$, *for all* $x \in \mathcal{S}$;
3. Transitivity: $x \leq y$ *and* $y \leq z$ *implies* $x \leq z$, *for all* $x, y, z \in \mathcal{S}$;
4. Antisymmetry: $x \leq y$ *and* $y \leq x$ *implies* $x = y$, *for all* $x, y \in \mathcal{S}$.

---

[3] A binary relation of a set $\mathcal{S}$ is a subset of the Cartesian product $\mathcal{S} \times \mathcal{S}$. Connexity simply means that all pairs are in the binary relation, one way or the other.

Given a total order, a strict total order[4] is the binary relation "<" defined by the condition

$$x < y \text{ if } x \le y \text{ and } x \ne y.$$

Gloss over this definition if you will, a total order and strict total order are simply a generalization of the *less than or equals to* and *less than* operators that act over real numbers ($\pi^2 < 10$), except now they act over elements of a set (dog < cat).

Definition 6.25. *A* categorical variable *is a variable that can take on any value from a finite set* C. *A categorical variable is said to be* ordinal *if its underlying set has a natural total order, otherwise it is said to be* nominal.

Note 6.8. We can always contrive an arbitrary ordering for any set, which is why we require ordinal variables to possess a *natural* total order; i.e., a total order that has meaning with respect to the meaning of the variable.
▷

Note 6.9. An ordinal feature differs from a numeric feature as it possesses no sense of scale. For example, we might know that $A < B < C$, but the nature of an ordinal variable is that we have no way to determine *how much greater* one variable is to another.   ▷

Thus, an ordinal feature is simply a discrete feature that has a natural ordering. Examples include T-shirt size ($XS < S < M < L < XL$) and responses on a survey ("very dissatisfied" < "dissatisfied" < "neutral" < "satisfied" < "very satisfied"). Similarly, a nominal feature does not have such an intrinsic ordering. Examples of nominal features include color (red, blue, yellow) and country (UK, CA, JP, AU).

The presence or lack of ordering affects how we preprocess a discrete feature.

Ordinal Features

In order to make ordinal features suitable for learning algorithms, we use a technique referred to as *label encoding*.

Definition 6.26. *Let* $X \in C$ *be an ordinal variable. Then the mapping* $\iota : C \to \{1, \ldots, |C|\}$ *is an* ordinal encoding *if it is invariant with respect to the strict total ordering on* C; *i.e., if*

$$x < y \text{ implies } \iota(x) < \iota(y), \text{ for all } x, y \in C.$$

---

[4] A strict total order satisfies the same set of axioms as a total order, except instead of reflexivity it is irreflexive, $x \not< x$ for all $x \in S$, and instead of antisymmetry it is asymmetric, $x < y$ implies $y \not< x$ for all $x, y \in S$.

Note 6.10. The comparison $x < y$ in Definition 6.26 is with respect to the ordering of the set $\mathcal{C}$, whereas the comparison $\iota(x) < \iota(y)$ is with respect to the standard Cartesian ordering of the integers.   ▷

Example 6.5. Fortunately, the scikit-learn (sklearn) package[5] has a built-in ordinal encoder, ready for use. In Code Block 6.1, we define three ordinal variables: survey, t-shirt, and color. We then instantiate an instance of the OrdinalEncoder class on line 7, declaring the required ordering of each feature as well as how to handle unknown values. The encoder fits the data frame (which is required, despite being, in this case, redundant from the object initialization) on line 17 and transforms it into an encoded matrix on line 18. Note that since we are using an ordinal encoding, we are actually saying that red < yellow < blue, at least for the purpose of the current illustration.

```
1  from sklearn.preprocessing import OrdinalEncoder
2
3  survey = ['very dissatisfied', 'dissatisfied', 'neutral',
       'satisfied', 'very satisfied']
4  t_shirt = ['XS', 'S', 'M', 'L', 'XL']
5  color = ['red', 'yellow', 'blue']
6
7  enc = OrdinalEncoder(
8         categories = [survey, t_shirt, color],
9         handle_unknown='use_encoded_value',
10        unknown_value=-1)
11
12 df = DataFrame({
13     'survey': np.random.choice(survey, size=10),
14     't_shirt': np.random.choice(t_shirt, size=10),
15     'color': np.random.choice(color, size=10)})
16
17 enc.fit(df)
18 X = enc.transform(df)
19 enc.categories_
```

Code Block 6.1: Syntax for sklearn's ordinal encoder

For this example, we generated our data frame by randomly choosing (with replacement) from our possible classes for each feature. A randomized data frame and its encoding is given in Table 6.4.   ▷

---

[5] For an up-to-date guide on using the scikit-learn package, see https://scikit-learn.org/stable/user_guide.html.

row	survey	t-shirt	color	X[:, 0]	X[:, 1]	X[:, 2]
0	satisfied	L	red	3	3	0
1	very dissatisfied	XS	red	0	0	0
2	very dissatisfied	M	red	0	2	0
3	satisfied	L	yellow	3	3	1
4	satisfied	S	blue	3	1	2
5	satisfied	L	blue	3	3	2
6	neutral	L	red	2	3	0
7	very satisfied	S	yellow	4	1	1
8	very dissatisfied	XL	yellow	0	4	1
9	very satisfied	M	yellow	4	2	1

Table 6.4: Example of an ordinal encoding.

Note 6.11. In a multi-class classification problem, one must further encode the target variable. This is achieved with a *label encoder*, available in sklearn's `LabelEncoder` class. The label encoder functions almost like the ordinal encoder, with a few exceptions:

- The label encoder does not accept any arguments into its constructor (except for, optionally, specifying the number of classes),
- The fit method may only be applied to a one-dimensional array or a pandas Series (equivalently, a single column of a data frame).

The label encoder is specifically used for encoding a *dependent* (i.e., target) variable.   ▷

Nominal Features

Similar to how we encoded ordinal features into integers, we encode nominal categorical features into binary vectors.

Definition 6.27. *Let* $X \in C$ *be a nominal categorical variable. Then the mapping* $\iota : C \to \mathbb{B}^{|C|}$ *is a one-hot encoding if it is an bijection (i.e., if it is one-to-one and onto) and if it satisfies the normalization condition*

$$||\iota(x)||_1 = 1, \ for \ all \ x \in C,$$

where $|| \cdot ||_1$ *is the* ℓ_1 *norm[6] and the set* $\mathbb{B} = \{0, 1\}$ *is the set of binary digits (bits).*

Note 6.12. The normalization condition in Definition 6.27 implies that the encoding $\iota(x)$ is a binary vector consisting of all zeros except for a solitary one.   ▷

---

[6] Here, the quantity $||x||_1 = \sum_{i=1}^{n} |x_i|$ represents the ℓ_1 *norm*, or *Manhattan norm*, of the vector $x \in \mathbb{R}^n$.

Example 6.6. Consider the set of colors $\mathcal{C} = \{R, B, Y\}$. The mapping defined by

$$\iota(R) = \begin{bmatrix} 1 \\ 0 \\ 0 \end{bmatrix}, \qquad \iota(B) = \begin{bmatrix} 0 \\ 1 \\ 0 \end{bmatrix}, \qquad \iota(Y) = \begin{bmatrix} 0 \\ 0 \\ 1 \end{bmatrix},$$

constitutes a one-hot encoding.   ▷

Example 6.7. Let's return to Example 6.5, but treat the variables as nominal, and implement a one-hot encoding. This is achieved in Code Block 6.2 (using the dataframe definition from Code Block 6.1). Here, we squash the fit and transform methods into a single command **fit_transform**, which is available to both encoders. Note, we pass in **sparse=False** if we actually want to visually see or print the output array X. Otherwise, the transform method will return a sparse array object, which is computationally more efficient for large data sets.

```
from sklearn.preprocessing import OneHotEncoder

enc = OneHotEncoder(sparse=False)
X = enc.fit_transform(df)
enc.categories_
```

Code Block 6.2: Syntax for sklearn's one-hot encoder

neut.	sat.	vy. dissat.	vy. sat.	L	M	S	XL	XS	blue	red	yellow
0	1	0	0	1	0	0	0	0	0	1	0
0	0	1	0	0	0	0	0	1	0	1	0
0	0	1	0	0	1	0	0	0	0	1	0
0	1	0	0	1	0	0	0	0	0	0	1
0	1	0	0	0	0	1	0	0	1	0	0
0	1	0	0	1	0	0	0	0	1	0	0
1	0	0	0	1	0	0	0	0	0	1	0
0	0	0	1	0	0	1	0	0	0	0	1
0	0	1	0	0	0	0	1	0	0	0	1
0	0	0	1	0	1	0	0	0	0	0	1

Table 6.5: Example of a one-hot encoding.

The resultant matrix is shown in Table 6.5. Now, the actual output matrix X is just a two-dimensional numpy array. In order to determine the column headings, we have to look at the attribute **enc.categories_**.

Notice that the columns do not follow the same ordering that we specified in Example 6.5. We could have passed in a `categories` argument into the constructor to have specified the ordering of the encoding. However, this is tantamount to shuffling the columns and does not typically make a difference in practice, as it does with ordinal encoding, where the relative size matters.

Notice, also, that there is not a column for *dissatisfied*, as that category does not occur in our example data frame, from Table 6.4. The `fit` method, therefore, does not create space for this feature. To include *dissatisfied*, the categories must be passed into the constructor, as in Code Block 6.1.   ▷

Even though a one-hot encoding returns a much larger matrix, which requires a separate column for each category within each nominal feature, it is necessary when encoding nominal data, i.e., data without any intrinsic ordering. Use of an ordinal encoder on nominal data, as so often with youth, leads to folly, as a learning algorithm will treat the encoded data as ordered.

6.2.2 Continuous Features

Continuous features are features which can take any real value. Unlike categorical features, they are already numeric. However, they often require their own form of manipulation prior to feeding into a learning algorithm. As was the case with categorical features, many of those transformations are built in to scikit-learn.

Discretization

Our first method for continuous features is a method that can be used to convert a continuous feature into a categorical feature. The quintessential example is age range: the raw age of a customer is converted into an ordinal age-range feature: 18–14, 25–34, 35–44, 45–54, 55–64, 65+. Instead of working with a customer's raw age, which, as an integer, is equivalent to dealing with an ordinal feature with $O(100)$ values, we work with the age-range, which is a more manageable set of seven buckets. In addition, there is usually an eighth bucket for *unknown* values.

Definition 6.28. *A mapping $\iota : \mathbb{R} \to \mathcal{C}$ shall constitute a* discretization *or* binning *of the real numbers if $\mathcal{C}$ is a finite, totally ordered set and if the mapping ι is invariant with respect to the total ordering, i.e., if*

$$x \leq y \text{ if and only if } \iota(x) \leq \iota(y).$$

A discretization is an ordinal discretization *if the set $\mathcal{C}$ is ordinally encoded, and it is a* one-hot discretization *if the set $\mathcal{C}$ is one-hot encoded.*

Typically, when we speak about discretization, we will implicitly refer to ordinal discretization, as the real numbers have a natural ordering.

The `sklearn.preprocessing` package has a built-in class, `KBinsDiscretizer`, that can handle most of our discretization needs for us. The constructor of this class has the following keyword arguments:

1. *n_bins* (int, default=5): The number of bins to use, must be an integer greater than 1.
2. *encode* ('onehot' (default), 'onehot-dense', 'ordinal'): Determines whether to return an ordinal encoding, or a one-hot encoding, which can be either dense or sparse.
3. *strategy* ('uniform', 'quantile' (default), 'kmeans'): Determines how the widths of the bins are calculated: uniform strategy will produce bins with equal widths; a quantile strategy will bin the data points (approximately) equally into bins; and a kmeans strategy will implement a clustering technique known as k-means, which we will not discuss at the present.

As with the OrdinalEncoder, LabelEncoder, and OneHotEncoder classes, the KBinsDiscretizer has a `fit`, `transform`, and a `fit_transform` method. The KBinsDiscretizer can discretize multiple numeric features at once; in this case, an object from this class should be instantiated with an array of integers passed in for the **n_bins** argument. Attributes of this class include **n_bins_**, an array with the number of bins for each feature, and **bin_edges_**, an array of arrays that determines the edges of the bins for each feature.

Example 6.8. Consider student data generated from a large university course: the student's age, exam score, and attendance percentage. The data are generated in Code Block 6.3. Age is then encoded into seven bins, exam score into twelve bins, and attendance into five bins. A quantile strategy is used, so that approximately an equivalent number of data points should appear in each bucket.

We can view the bin edges and the first few rows of our data frame and output matrix, as shown in lines 14–15.   ▷

Scaling Numeric Features

Many machine learning algorithms are highly sensitive to the scaling of numeric features and perform much better when the feature set is appropriately scaled. For example, if an algorithm were to use the Euclidean distance between two numeric features for the purpose of predicting whether or not a home was in a good school district, we should expect things to break down if one feature is the number of bedrooms and another is the home price: these two features live on radically different scales.

The `sklearn.preprocessing` package has three main types of built-in scalers that represent the three most commonly used forms of scaling.

```
1   from sklearn.preprocessing import KBinsDiscretizer
2
3   n_rows = 100
4   df = DataFrame({
5           'age': 18 + np.random.exponential(10,
                  size=n_rows).astype(int),
6           'score': np.fmin(100, np.random.normal(loc=70, scale=20,
                  size=n_rows).astype(int)),
7           'attendance': np.random.beta(5, 1, size=n_rows)
8               })
9
10  K = KBinsDiscretizer(n_bins=[7, 12, 5], encode='ordinal',
        strategy='quantile')
11  X = K.fit_transform(df)
12  # K.fit(df)
13  # X = K.transform(df)
14  print(K.bin_edges_)
15  print(df.head(10), X[:10, :])
```

Code Block 6.3: Syntax for sklearn's discretizer

Definition 6.29. *Let* $X_1, \ldots, X_n \sim f_X$ *be a sample of* IID *data. Then the transformed data set*

$$Z_i = \frac{X_i - \overline{X}_n}{S_n},$$

where $\overline{X}_n$ *is the sample mean and* S_n^2 *is the sample variance, is said to be a* z-score normalized *or* standardized *data set. Similarly, the transformed data set*

$$Y_i = \frac{X_i - \min(X_i)}{\max(X_i) - \min(X_i)}$$

is said to be a min-max normalized *data set. Finally, the transformed data set*

$$W_i = \frac{X_i - Q_{0.5}}{Q_{1-\alpha} - Q_\alpha},$$

where $\alpha \in (0, 1/2)$, *typically* $\alpha = 0.2$ *or* $\alpha = 0.25$, *and* Q_α *is the* αth *quantile of the sample, is said to be a* quantile normalized *data set. The value* $Q_{0.5}$ *is the median of the original data set. For the case* $\alpha = 0.25$, *the denominator* $Q_{0.75} - Q_{0.25}$ *is referred to as the* interquartile range.

Common methods available in the **sklearn.preprocessing** module are

- **StandardScaler**: standardization produces scaled data with zero mean and unit variance;
- **MinMaxScaler**: min-max normalization produces scaled data on range $[0, 1]$;

- MaxAbsScaler: divides by the largest absolute value of the data, without shifting; scaled data on range $[-1, 1]$. Can be used with sparse data.
- RobustScaler: quantile normalization does not limit range.

The StandardScaler takes arguments with_mean and with_std (default True). If used with sparse data, with_mean should be set to False. The Min-MaxScalar takes arguments feature_range (default $(0, 1)$) and clip (default False). If clip is set to True, test data (not used during the original fit) is clipped to the same range $[0, 1]$ as the input data. Finally, the RobustScaler has arguments with_centering (default True), with_scaling (default True), and quantile_range (default $(0.25, 0.75)$).

Once an object of any of these classes is constructed, the standard fit, transform, and fit_transform methods are available to transform a data set.

In addition to the aforementioned scalers, scikit-learn has two nonlinear transformation methods, QuantileTransformer and PowerTransformer, which transform the data set into a uniform distribution or a Gaussian distribution, respectively.

6.2.3 Data Pipelines

In building a model with a large feature set, one typically encounters multiple transformations prior to training a model. This process will be simplified with a structure known as a pipeline. But first, we will consider two additional preprocessing tasks: handling missing values and randomizing the data into training and test sets.

Imputing missing values

Data sets often contain missing values. Instead of deleting an entire datum, we can define a strategy for handling missing data. To achieve this, we will use the SimpleImputer from the sklearn.impute package. The constructor has an argument strategy, which can take a value from 'mean,' 'median,' 'most_frequent,' and 'constant.' If 'constant' is selected, the argument fill_value should also be specified.

In Code Block 6.4, we define an outer join of two data frames, which results in two columns (b and c) with missing values. We then demonstrate imputing with the median and most frequent strategies. Note that the *mean* and *median* strategy are not available for categorical features.

Often, in practice, it is useful to use the constant strategy, in order to give a unique and separate value to the missing data.

Building a Pipeline

So far, we have reviewed many types of transformers: encoders, discretizers, scalers, and imputers. Each of these classes has a fit and a transform

```
1  from sklearn.impute import SimpleImputer
2
3  df_1 = DataFrame({'a':[1,2,5,6], 'b':['x','y','z','z']})
4  df_2 = DataFrame({'a':[0, 2, 3, 4, 6], 'c':[4, 8, 15, 16, 23]})
5  df = pd.merge(df_1, df_2, how='outer')
6
7  imp = SimpleImputer(strategy='median')
8  X_num = imp.fit_transform(df[['a', 'c']])
9  imp = SimpleImputer(strategy='most_frequent')
10 X_cat = imp.fit_transform(df[['b']])
11 imp = SimpleImputer(strategy='constant', fill_value='MISSING')
12 X_cat = imp.fit_transform(df[['b']])
```

Code Block 6.4: Imputing missing values

method, along with, for convenience a `fit_transform` method, which is equivalent to applying both methods to the same data set in order. The actual machine learning machinery lives in classes as well, a collection of classes known as *estimators*. Estimators typically have both a *fit* and a *predict* method, which operate separately: `fit` is used with the training data, and `predict` is used with the test data.

A *pipeline* is a series of transformers with an optional estimator in its last position. The `fit` method only need be called once on a pipeline object: The dataset is passed through the `fit_transform` methods of each transformer in series. The pipeline object will have all the functionality as its last estimator or transformer.

A pipeline is constructed by passing a list of key-value pairs: the key being an arbitrary (but useful) name for each step of the pipeline, and the value being the transformer or estimator object that is to be called at that step.

A simple pipeline is constructed in Code Block 6.5. The `mask` method on line 6 randomly (with 10% probability) *masks* or removes each value of the data frame. The pipeline consists of two transformers: impute missing values with the median, and then perform standardization.

A similar pipeline for categorical data is constructed in Code Block 6.6. Here, the pipeline consists of three steps: impute missing values, apply an ordinal encoder, apply a one-hot encoding. The application of a one-hot encoding following an ordinal encoding is completely redundant, and could have (and should have) been accomplished with the one-hot encoding alone. The redundancy was only for the purpose of illustration.

In practice, the pipelines in Code Blocks 6.5 and 6.6 would contain an estimator at the end of the pipeline, so that a model could have been built. We will see examples of this soon enough.

```
1  from sklearn.pipeline import Pipeline
2  from sklearn.preprocessing import StandardScaler
3
4  X = np.random.randint(0, 20, size=(10, 4))
5  df = DataFrame(X, columns=['a','b','c','d'])
6  df = df.mask(np.random.random(df.shape) < .1)
7
8  transformers = [
9        ('impute', SimpleImputer(strategy='median')),
10       ('standardize', StandardScaler())]
11
12 pipe = Pipeline(transformers)
13 pipe.fit_transform(df)
```

Code Block 6.5: A simple pipeline for continuous features

```
1  s = """Our revels now are ended. These our actors,
2  As I foretold you, were all spirits and
3  Are melted into air, into thin air"""
4
5  s = s.replace(",", "").replace(".","").replace("\n"," ").split(' ')
6  df = DataFrame(np.random.choice(s, size=(10,4)), columns=['a', 'b',
        'c', 'd'])
7  df = df.mask(np.random.random(df.shape) < .1)
8
9  transformers = [
10       ('impute', SimpleImputer(strategy='constant',
            fill_value='UNKNOWN')),
11       ('ordinal',
            OrdinalEncoder(handle_unknown='use_encoded_value',
            unknown_value=-1)),
12       ('ohe', OneHotEncoder(sparse=False,
            handle_unknown='ignore'))]
13
14 pipe = Pipeline(transformers)
15 pipe.fit_transform(df)
```

Code Block 6.6: A simple pipeline for categorical features

The astute reader may be wondering how to handle a feature set that
contains mixed data types. This, too, is easily accomplished with the construction of a `ColumnTransformer` object, which can then be inserted as
a pipe in the pipeline. The column transformer simply specifies which
transformations to apply to which columns. Finally, a keyword argument
`remainder` tells the transformer what to do with columns not specified:
'drop' (default) or 'passthrough'. An example pipeline using a column transformer is constructed in Code Block 6.7.

```python
from sklearn.compose import ColumnTransformer

class_levels = ['freshman', 'sophomore', 'junior', 'senior']
majors = ['science', 'engineering', 'math', 'computer science']

n_rows = 100
df = DataFrame({
        'level': np.random.choice(class_levels, size=n_rows),
        'major': np.random.choice(majors, size=n_rows),
        'age': 18 + np.random.exponential(10,
            size=n_rows).astype(int),
        'score': np.fmin(100, np.random.normal(loc=70, scale=20,
            size=n_rows).astype(int)),
        'attendance': np.random.beta(5, 1, size=n_rows)})

column_transformer = ColumnTransformer(
        [('ordinal', OrdinalEncoder(categories=[class_levels],
            handle_unknown='use_encoded_value', unknown_value=-1),
            ['level']),
         ('ohe', OneHotEncoder(sparse=False,
            handle_unknown='ignore'), ['major']),
         ('bins', KBinsDiscretizer(n_bins=8, encode='ordinal',
            strategy='quantile'), ['age', 'score'])],
        remainder='passthrough')

column_transformer.fit(df)
```

Code Block 6.7: A column transformer can handle mixed data types

Now, ideally, you get some ideas. For example, following the illustration
in Code Block 6.7, we might think to ourselves that we should construct two
column transformers—one to impute missing values and one to apply the
various encodings—and connect them together in a pipeline. A problem
with this is that the `transform` method outputs an *array*, not a pandas
data frame.

To remedy this, I devised a clever workaround. The idea is we can define our own subclass of `ColumnTransformer` that returns a pandas data frame from its `transform` and `fit_transform` methods, instead of an array. The code is given in Code Block 6.8. A final caveat: the `OrdinalEncoder` does not seem to actually handle unknowns as advertised, which requires us to redefine the `class_levels` list to include our unknown value. (It is important to actually redefine line 22 with the updated list of `class_levels`.) Finally, this method only works if *every* column is explicitly handled in the column imputer; hence the use of `remainder='drop'`.

```python
class ColumnImputer(ColumnTransformer):
    def transform(self, X):
        X = ColumnTransformer.transform(self, X)
        cols = []
        for t in self.transformers_:
            cols += t[2]
        return DataFrame(X, columns=cols)

    def fit_transform(self, X, y=None):
        X = ColumnTransformer.fit_transform(self, X, y)
        cols = []
        for t in self.transformers_:
            cols += t[2]
        return DataFrame(X, columns=cols)

column_imputer = ColumnImputer([
        ('impute_cat', SimpleImputer(strategy='constant',
            fill_value='UNKNOWN'), ['level', 'major']),
        ('impute_num', SimpleImputer(strategy='median'), ['age',
            'score', 'attendance']),
        remainder='drop'])

class_levels = ['UNKNOWN', 'freshman', 'sophomore', 'junior',
    'senior']
column_transformer = ColumnTransformer( # ... Same as before.

pipe = Pipeline([
        ('impute', column_imputer),
        ('encode', column_transformer)])

pipe.fit_transform(df)
```

Code Block 6.8: Multiple column transformers in a pipeline; continuation of Code Block 6.7.

Training and Test Sets

Naturally, scikit-learn also has a built-in tool for handling the randomization of a data set (features and labels) into a training and test set. The syntax is shown in Code Block 6.9.

```
from sklearn.model_selection import train_test_split
from sklearn.metrics import r2_score

X = np.random.randint(0, 20, size=(100, 4))
y = np.random.randint(0, 2, size=100)
X_train, X_test, y_train, y_test = train_test_split(X, y,
    test_size=0.20, random_state=42)

pipe = Pipeline() #.....
pipe.fit(X_train, y_train)
y_pred = pipe.predict(X_test)
print(f"R2 score: {r2_score(y_test, y_pred)}")
```

Code Block 6.9: Train-test split

Note 6.13. In Code Block 6.9, notice that `train_test_split` is used directly without being instantiated. This is because it is a function, not a class. ▷

Here, we suppose that we have built a pipeline with the various preprocessing operations, topped with an actual model, or estimator, with a `predict` method. The code shows how we can automatically create a train-test split, train the model on the training data, and use the holdout data for our validation step.

6.3 Object-Oriented Data Science

Object-oriented programming (OOP) is a software-engineering paradigm centered around the use of modular, reusable code. OOP has allowed engineers to develop sophisticated software architectures and platforms that would otherwise be intractable. Though much of this text is focused around the statistical aspects of data science, we devote this section to some of the engineering aspects. We advocate for object-oriented design in the development of data-science projects in order to both better leverage the reusability aspects and to better manage sophisticated projects that are developed across a team. We conclude this section with a discussion of *agile*, which is a process for managing engineering projects that was developed around a set of principles known as the *agile manifesto*.

For more in depth introduction to data structures in Python, see Lambert [2014] and Lee and Hubbard [2015]. For more on data structures and algorithms, see Lafore [2003] (JAVA) or Cormon *et al.* [2009].

6.3.1 Classes and Objects

Object-oriented programming is focused on the use of *classes* and their specific realizations, *objects.* The class encapsulates a set of instructions for creating (or instantiating) individual objects. The class is the cookie cutter and the objects are the cookies.

Definition 6.30. *In* object-oriented programming, *a* class *is a code block that consists of a number of functions, called* methods, *and variables, called* attributes, *that provides instructions for how to construct, or* instantiate *any number of* objects. *An* object (or, instance of a class) *is a variable that is created from a class, that stores all of the class's methods and attributes. Two instances (objects) from the same class may have different values for their attributes, but maintain the same structure, as defined by the underlying class.*

In Python, the instructions for instantiating an object from a class are given by the __init__ method, which is reserved for such use. This method is not called directly, rather we use the name of the class like a function that takes in the prescribed input parameters and returns an object of that class. In addition, all methods of a class must take at least one argument, called **self**, which represents the object itself. It is through this argument that any method from a class can access the values stored as attributes and any of the other methods for any particular object. To illustrate, consider the following.

Example 6.9. Consider the class defined in Code Block 6.10, which defines a concept of a *car.*

Each "car object" that is constructed from this class has six attributes: name, color, max speed, position, velocity, and time. The values of those attributes will differ from car to car, though the structure itself remains constant. In addition, each object will possess four methods: speed up, time lapse, stop, and repaint.

Individual cars may be instantiated using the __init__ method, though this method is not called directly. Two cars are instantiated in Code Block 6.11. In addition, we provide a method called __str__ which returns a string representation of an individual object[7]. This is used with the **print** command, as shown in Code Block 6.11.

In Code Block 6.11, the objects are represented by the variables car_1 and car_2. In addition to printing the object, as defined by the internal

---

[7] It is not common to define a __str__ method for data-science applications, but it seemed helpful for this example

```python
class Car:
    def __init__(self, name, color, max_speed=100):
        self.name = name
        self.color = color
        self.max_speed = max_speed

        self.position = 0
        self.velocity = 0
        self.time = 0

    def __str__(self):
        s  = f"name: {self.name}\n"
        s += f"color: {self.color}\n"
        s += f"max speed: {self.max_speed}\n"
        s += f"position: {self.position}\n"
        s += f"velocity: {self.velocity}\n"
        s += f"time: {self.time}\n"
        return s

    def speedUp(self, delta_v=0):
        self.velocity = min(self.max_speed, self.velocity + delta_v)

    def timeLapse(self, t):
        self.time += t
        self.position += t * self.velocity

    def stop(self):
        self.velocity = 0

    def repaint(self, new_color):
        self.color = new_color
```

Code Block 6.10: Car class for Example 6.9

__str__ method defined in the class, we may also print specific attributes using, for example, print(car_1.speed). Finally, we may modify any of the attributes directly, using, for example, car_1.speed += 200.   ▷

6.3.2 Principles of OOP

Object-oriented programming is built around the following four principles:

1. *Inheritance*: We may define new classes ("children") from old ("parents") by updating only the new aspects, whereas all functionality not specifically modified is *inherited* from the parent class;
2. *Encapsulation*: An object only exposes certain methods and attributes to the outside world, while keeping unnecessary details hidden;

```
1   # Instantiate two car objects
2   car_1 = Car('1GAT123', 'blue', max_speed=140)
3   car_2 = Car('EIPI+1', 'red', max_speed=160)
4
5   car_1.speedUp(100)
6   car_1.timeLapse(3)
7   car_2.speedUp(80)
8   car_2.timeLapse(3)
9   car_1.stop()
10  car_1.repaint('black')
11  print(car_1) # invokes __string__ method
12  print(car_2) # invokes __string__ method
13  print(car_1.speed) # access particular attribute
14  car_1.speed += 200 # modify attribute directly
```

Code Block 6.11: Creating objects from the Car class

3. *Abstraction*: The operations of an object can be defined without reference to its internal implementation;
4. *Polymorphism*: Code can be agnostic with respect to which class it is operating on, as long as it is operating from a given inheritance hierarchy; similarly, two classes defined within an inheritance may have different internal definitions for the same method.

While encapsulation and abstraction both hide data, they differ in a crucial manner: encapsulation hides data at an implementation level, whereas abstraction hides data at a design level. We discuss each of these four principles in turn, focusing on how they are implemented in Python.

Inheritance

Inheritance allows us to define new classes from old, changing only what is needed. In this context, the new class is called a *child* of the old class, which is called the *parent*. For example, our **Car** classof Example 6.9 might be a subclass from a **Vehicle** class. The **Vehicle** class would define functionality and attributes common to all kinds of vehicles: bicycles, trains, boats, cars, planes, and so forth. The **Car** class could then overwrite—or freshly define—those elements that are unique to automobiles. This process can continue: the **Car** class might then be a parent to a number of subclasses representing different makes of cars (BMW, Subaru, Ford, etc.). Perhaps the **Car** class has a **engine** method, which could be uniquely defined for each of its subclasses.

In addition, we can allow for *multiple inheritance*, in which a new class is defined from an ordered sequence of parents. This can lead to ambiguity if one is not careful, as demonstrated by the *diamond problem* of Figure 6.4.

In this figure, class D inherits from both classes B and C, which are each,

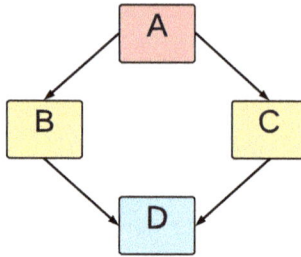

Fig. 6.4: Diamond problem: Class D inherits from both B and C.

in turn, children of the class A. In Python 3, the method resolution order is $D \to B \to C \to A$. (This differs from Python 2, which used a DLR—depth-first left-to-right—method, which would have favored A over C.)

In other words, suppose that class A has a method, which is overwritten by class C, but not by class B or by class D. Objects of type D would then follow the overwritten method as defined by class C.

Example 6.10. We can construct a subclass from the **Car** class (Code Block 6.10) that provides for a more realistic **speedUp** method, in the sense that it takes the vehicle's acceleration into account, so that the velocity jump is not instantaneous. The result is shown in Code Block 6.12.

```
class RealCar(Car):
    def __init__(self, name, color, max_speed=100, acc=5):

        self.acc = acc
        super().__init__(name, color, max_speed=max_speed)

    def speedUp(self, delta_v=0):
        delta_v = min(delta_v, self.max_speed - self.velocity)
        delta_v = max(-self.velocity, delta_v)
        delta_t = abs(delta_v) / self.acc
        self.velocity += delta_v
        self.timeLapse(delta_t)
```

Code Block 6.12: Subclass of **Car**

Notice that we indicate the parent class in line 1. All methods not explicitly redefined within this code are inherited from the parent class, so that we may still access the **timeLapse**, **stop**, and **repaint** methods.

In our `RealCar` class, we update the `__init__` method, to allow for an additional input parameter, representing the vehicle's acceleration. We access the parent class using Python's built-in `super()` method. Finally, we update the `speedUp` method of the parent class with the code shown on lines 7–12.   ▷

Encapsulation

The concept of *encapsulation* refers to the art of only providing access to certain functionality to the outside world. Many programming languages have a concept of *private* and *public* attributes and methods: public attributes and methods are exposed to the outside world, whereas private attributes and methods are only available from within the class itself. For example, when building a car, there are many operations that are required for it to run. Only a handful of these methods are exposed to the driver: gas pedal, brakes, steering wheel. Even though a car requires many other operations to run—piston and cylinders, spark ignition, diverting energy to battery, and so forth—those functions are not directly exposed to the driver.

Python does not offer private attributes and methods. Instead, we may follow the convention that attributes and methods not intended to be exposed to the outside world be preceded by an underscore. For instance, we could signify our intent of keeping `timeLapse` as an internal, private method by renaming it as `_timeLapse`.

For stronger protection of private attributes and methods, we may instead use a double underscore; for example, `__timeLapse`. Attributes and methods with names beginning with a double underscore are not *directly* available from the outside, but need to be preceded by an additional underscore and the class name. For instance, to access `__timeLapse` from *outside* of the class, we first instantiate an object `my_car = Car(..)`, and then we call `my_car._Car__timeLapse(5)`. This process is known as *name mangling*, and is a layer of protection Python offers for variables intended to be kept private.

When defining classes within an inheritance hierarchy, we take the following approach: for private attributes and methods meant to be accessible to (or overwritten by) the various children and descendants, we use the single underscore, whereas private attributes and methods that are meant to be attached to a single class, we use the double underscore. The double underscore mangles the variable name when referred to from any subclass, making it more difficult to access (or overwrite) the original functionality.

Abstraction

Whereas encapsulation hides specific chunks of code or data from the outside world, at the implementation level, abstraction hides the details of an implementation at the design level. In other words, abstraction is used to define the inputs and outputs without any reference to any specific implementation. In this way, abstraction defines the interface that any user of a class must follow. Such an interface is often referred to as an *application user interface*, or *API*.

In practice, abstraction is used whenever we want to define a family of classes that follow a common API. We achieve this by constructing an *abstract class*, which contains one or more *abstract methods*, which are simply methods that have not been implemented. This defines a set of methods, along with their inputs and outputs, that must be defined for any subclass of the abstract class. Abstract classes can be written in Python using the abc (abstract base class) package, as shown in Code Block 6.13.

```python
from abc import ABC, abstractmethod

class AbstractVehicle(ABC):

    @abstractmethod
    def speedUp(self, delta_v: int) -> None:
        pass

    @abstractmethod
    def timeLapse(self, t: float) -> None:
        pass

    @abstractmethod
    def stop(self) -> None:
        pass
```

Code Block 6.13: Abstract classes and methods in Python.

Thus, to create an abstract base class in Python, we simply subclass the ABC class from the abc package. To define abstract methods, we *decorate* the method by adding the text @abstractmethod, which is referred to as a *decorator*, to the line prior to the method definition. Decorators are also known as *wrappers*, as they can be used to wrap our functions in a blanket of code. We will discuss wrapper functions and decorators shortly, but for now, think of the extra line as a magic spell that prevents the class, or any of its subclasses, from instantiating objects unless they have overridden the abstract methods with specific implementations.

Code Block 6.13 defines three methods that must be common to all vehicles: speedUp, timeLapse, and stop. Since they are abstract methods, Python will throw an error if we try to instantiate an object from the AbstractVehicle class. However, when we defined Car, in Code Block 6.10, we could replace the first line with

```
class Car(AbstractVehicle):
```

to ensure that our `Car` class satisfies the API defined by our concept of an abstract vehicle. We have also established a lengthy hierarchy of subclasses:

$$ABC \rightarrow \text{AbstractVehicle} \rightarrow \text{Car} \rightarrow \text{RealCar}.$$

It's not hard to imagine a hierarchical tree structure: we might have many types of vehicles (cars, boats, airplanes, submarines), which each might have many types of manifestations (car alone might branch to BMW, Mercedes, Subaru, Ford, Ferrari, Toyota, Trabant, etc.).

Finally, we used *type hinting* to specify the types of inputs and outputs we should expect from our abstract methods. Type hinting is not limited to abstract classes; it can be used when defining methods for any class. We typically omit type hinting from our examples to save space, though it is typically beneficial to use in practice.

Polymorphism

Our final principle of OOP is polymorphism, which simply means that objects from different classes can be designed to share behaviors. For example, all vehicles descendant from the `AbstractVehicle` method defined in Code Block 6.13 must have an implementation for `speedUp`, `timeLapse`, and `stop`. Their internal workings might vary dramatically: how an airplane *speeds up* is quite different than how a skateboarder speeds up. Polymorphism allows us to instantiate a fleet of vehicles from our `AbstractVehicle` family—cars, bicycles, trains, planes, ships, submarines, and skateboarders—and use them agnostically in regards to their internal workings. We will return to this concept in our discussion of factory methods, and again when we discuss how the object-oriented paradigm can be used in a data-science context.

6.3.3 Generators, Wrappers, and Factories

Special Methods

Certain methods of a class have names that are both preceded and followed by double underscore, such as `__init__` and `__str__`. These are referred to as *special methods*, and each have specific meaning related to how objects can be used in a Python environment. Commonly used special methods are shown in Tables 6.6 and 6.7. For a more comprehensive list, see `docs.python.org`[8]. Though many of these special methods are not used in practice, it is useful to be aware of them.

---

[8] https://docs.python.org/3/reference/datamodel.html#special-method-names

method	usage	explanation
__init__(self,...)	x=X(...)	instantiates an object from class X
__str__(self)	print(x)	returns a string representation of object
__len__(self)	len(x)	returns the length of object
__eq__(self, other)	x==y	compares two objects for equality
__getitem__(self, key)	x[key]	gets the value associated with a key
__setitem__(self, key, value)	x[key] = value	sets value of a key
__contains__(self, item)	item in x	determine if item in object
__iter__(self)	iter(x)	returns an iterator object
__next__(self)	next(x)	returns next item from container

Table 6.6: Special methods in Python for class X with objects x and y.

method	usage	explanation
__add__(self, other)	x + y	add
__sub__(self, other)	x - y	subtract
__mul__(self, other)	x * y	multiply
__truediv__(self, other)	x / y	divide
__pow__(self, other)	x ** y	power
__lt__(self, other)	x < y	less than
__le__(self, other)	x <= y	less than or equal to
__eq__(self, other)	x == y	equals to
__ne__(self, other)	x != y	not equals to
__gt__(self, other)	x > y	greater than
__ge__(self, other)	x >= y	greater than or equal to

Table 6.7: Math-type special methods in Python for class X with objects x and y.

Iterators and Generator Functions

In python, an *iterable object* is simply an object that is an object that is capable of returning its data, one element at a time. Iterables are often used in the context of a `for` loop. An iterable object must implement two special methods, `__iter__` and `__next__`, collectively known as the *iterator protocol*. Lists, tuples, and dictionaries are all *iterable* objects.

An *iterator* is an object that represents a specific stream of data. The `__iter__` method of an iterable object must return an iterator. Once an iterator has been instantiated, repeated calls to its `__next__` method return successive items from the stream, until a `StopIteration` exception is raised.

For example, if we define a list `my_list = [1,2,3,4,5]`, calling `next(my_list)` throws an error, as `my_list` is not itself an iterator. Instead, if we call `my_iter = iter(my_list)`, we can then call `next(my_iter)` multiple times, receiving, in turn, the values 1–5, and the `StopIteration` exception thereafter.

Example 6.11. An iterable that represents the Fibonacci sequence is shown in Code Block 6.14.   Once an object has been instantiated, we can iterate

```python
class Fibonacci:
    def __init__(self, max_length=10):
        self.max_length = max_length
    def __iter__(self):
        self.position = 0
        self.lag1 = 0
        self.lag2 = 0
        return self
    def __next__(self):
        self.position += 1
        if self.position > self.max_length:
            raise StopIteration

        next_item = 1 if self.position == 1 else self.lag1 + self.lag2
        self.lag1, self.lag2 = next_item, self.lag1
        return next_item

fib = Fibonacci()
for x in fib:
    print(x) # prints 1, 1, 2, 3, 5, 8, 13, 21, 34, 55
```

Code Block 6.14: Defining an iterable in Python.

over it using a for-loop, as shown on lines 18–20.   ▷

The code required to define an iterator is, however, a bit verbose. A commonly used simplification is that of a *generator function*, which creates an iterator without the need to explicitly invoke the iterator protocol. Generator functions are similar to regular Python functions, except they have one or more **yield** statements, instead of the typical **return** statement.

Example 6.12. The Fibonacci numbers can be generated using the generator function of Code Block 6.15, achieving the same result as Code Block 6.14.

```python
def fibonacci(max_length=10):
    position = 0
    lag1 = 0
    lag2 = 0
    while position < max_length:
        position += 1
        next_item = 1 if position == 1 else lag1 + lag2
        lag1, lag2 = next_item, lag1
        yield next_item


for x in fibonacci():
    print(x) # prints 1, 1, 1, 3, 5, 8, 13, 21, 34, 55


f = fibonacci()
next(f) # 1
next(f) # 1
next(f) # 2
# .... 3, 5, and so forth...


sum(fibonacci()) # 143
```

Code Block 6.15: Fibonacci numbers using generators

The generator function **fibonacci** defined in Code Block 6.15 can be used in the context of a for loop, as shown on lines 11–12. The function itself returns a fresh iterator object, which can be used to get successive values through the **next** method, as shown on lines 14–17. Additionally, we can simply sum the iterator object returned by a call to **fibonacci()**, as shown on line 19.   ▷

Wrapper Functions and Decorators

In Python, a *wrapper function* is a special type of function that takes a function as input[9] and returns a new function as output. They can be

---

[9] In Python, functions are objects, and therefore they may be passed into other functions as inputs.

used to "wrap" around other functions by adding a decorator—a line of code with the @ symbol followed by the name of the wrapper function— to the line *preceding* the function definition, in the same way we used the @abstractmethod command to decorate various functions in Code Block 6.13. An example of a simple wrapper is shown in Code Block 6.16.

```python
def greetings(func):

    def wrapper(*args, **kwargs):
        print("Hello, there!")
        result = func(*args, **kwargs)
        print("Googdbye!")
        return result

    return wrapper

@greetings
def pow2(x, power=2):
    return x**power

t = pow2(5, power=3) # Prints Hello / Goodbye; sets t = 125
```

Code Block 6.16: A simple wrapper.

Wrappers are often used when defining hierarchies. For example, there might be a number of assertions, or data checks, that many methods of various subclasses share. Instead of writing these assertions into each individual method, we can use a single wrapper function in the parent class that is then inherited to each subclass. For example, we can define a wrapper in the AbstractVehicle abstract class defined in Code Block 6.13. The code is shown for defining the wrapper function is shown in lines 5–13 of Code Block 6.17.

The _timeWrap decorator can be used to wrap any of the methods from the Car class, which is subclassed from AbstractVehicle. Proper usage of the decorate is shown on line 19. Notice that in order to access the decorator, we must make reference to the class in which the method is defined. (Technically, the method _timeWrap is implemented as a static method, which we discuss next.)

Static Methods and Class Methods

Sometimes it is useful to define methods, within a class, that can be called *without reference to any particular object*. This scenario occurs with two distinct flavors: static methods and class methods. *Static methods* are simply

```
1   class AbstractVehicle(ABC):
2
3       # .... Methods from AbstractVehicle
4
5       def _timeWrap(func):
6
7           def wrapper(*args, **kwargs):
8               t = time.time()
9               results = func(*args, **kwargs)
10              print(f"Time to execute {func.__name__} is: {time.time()
                    - t}")
11              return results
12
13          return wrapper
14
15  class Car(AbstractVehicle):
16
17      # ....
18
19      @AbstractVehicle._timeWrap
20      def speedUp(self, delta_v=0):
21          # ....
```

Code Block 6.17: Adding a wrapper to the **AbstractVehicle** class.

methods attached to a class that can be used without instantiating an object from that class. *Class methods* are similar, except that a class method is allowed to modify any of the *class attributes*, or variables attached to the class itself, as opposed to individual objects. Class attributes are uniform and accessible across all objects from the class; this is a particularly handy way to store constant that do not vary object-to-object. Class methods must take the variable cls, instead of self, as their first argument, which refers to the class itself, and not the object. Static methods should have neither cls nor self as arguments, as they are pure functions attached to the class, and can modify neither object nor class.

Class methods should be preceded by the @classmethod decorator. Static methods should be preceded by the @staticmethod operator. The exception to the latter rule is when defining static decorators, as in lines 5–13 of Code Block 6.17. Though _timeWrap is a static method, the absence of the @staticmethod decorator simply means it is not available when attached to any object of the class.

Factories

Static methods are commonly used to define *factories*, which are certain methods used to generate new objects from a particular family of classes.

Example 6.13. Suppose that we have defined a variety of subclasses from the `AbstractVehicle` class, including `Bmw`, `Submarine`, and `Airplane`. A simple factory method and its usage is shown in Code Block 6.18.

```python
class VehicleFactory:

    @staticmethod
    def get(name, *args, **kwargs):
        lookup = {'bmw': Bmw,
                  'sub': Submarine,
                  'airplane': Airplane}

        return lookup[name](*args, **kwargs)

c = VehicleFactory.get('bmw', 'my sports car', 'red')
```

Code Block 6.18: Factory method to generate various objects from the `AbstractVehicle` family.

Note that, unlike other classes we've studied, the `VehicleFactory` never gets instantiated. We *directly* invoke the `get` method of the `VehicleFactory` class, without reference to any specific instance of the class. This method takes as argument a name, used as the key to a lookup map, and returns an instantiated object of the appropriate type.

We could, alternatively, define the lookup map *outside* of the `get` method, making it a class attribute. If we did so, we would instead define `get` as a class method, and reference `cls.lookup`.

Another variation is to use a factory method to *dynamically* generate new Python classes on the fly, using the built-in `type` function. Suppose, in addition to our `AbstractVehicle` class, we also had an `AbstractEngine` class, with various subclasses, such as `Diesel`, `Turbine`, `Combustion`, and `Rocket`. Further, suppose that any engine can be used for any type of vehicle, as they each follow a consistent API. By following the convention `vehicle_engine`, we can dynamically generate an object that inherits the vehicle components from the `vehicle` class and the engine components from the `engine` class, as shown in Code Block 6.19.

This factory returns an instantiated object from the new class, inheriting properties from both the vehicle as well as the engine. This is a convenient way to handle the case in which a class has multiple components,

```
 1   class VehicleFactory:
 2
 3       # class variables
 4       vehicles = {'bmw': BMW,
 5                   'sub': Submarine,
 6                   'airplane': Airplane}
 7
 8       engines = {'diesel': Diesel,
 9                  'turbine': Turbine,
10                  'combustion': Combustion,
11                  'rocket': Rocket}
12
13       @classmethod
14       def get(cls, name, *args, **kwargs):
15           assert '_' in name
16           vehicle, engine = name.split('_')
17
18           return type(name, (cls.vehicles[vehicle],
19               cls.engines[engine]), {})(*args, **kwargs)
20   c = VehicleFactory.get('bmw_rocket', 'my rocket car', 'red')
21   type(c) # prints: __main__.bmw_rocket
```

Code Block 6.19: Dynamic class creation in a factory method

and each component can have multiple variations. Instead of defining each permutation as an individual subclass, we can define an abstract class for each component, and dynamically cast one subclass from each component together to generate a new mix-and-match object.   ▷

6.3.4 OOP in Data Science

We next consider how we can deploy the principles of object-oriented programming to data science practice. In doing so, we seek to benefit from the various benefits offered by an object-oriented mindset, including use of modular, reusable code as well as a structure for building data-science products in teams.

By leveraging the four principles of object-oriented design, we can deploy greatly simplified runtime scripts, as shown in Code Blocks 6.20 and 6.21. Py3912 shows the overall structure of the file, whereas Code Block 6.20 defines the main function. Notice that all of the complexity is abstracted within three classes: our database class D, which can read and write to our database, our schema class S, which defines the table structure (e.g., column types, partitions, etc.) for each table we wish to write to, and our model class M, which encapsulates all of the implementation details of our

```python
def main(project, model, mode, db_src, exec_id, **config):

    D = DataBaseFactory.get(db_src)
    S = SchemaFactory.get(project)

    S.set('run_config')
    df_config = DataFrame({'project': project,
                           'model': model,
                           'mode': mode,
                           'exec_id': exec_id,
                           'config': str(config)}, index=[0])
    D.put(S, df_config)

    input_table = config.pop('input_table')
    target_table = config.pop('target_table')
    cols = config.pop('cols')
    target_cols = config.pop('target_cols')
    time_col = config.pop('time_col')
    point_in_time = config.pop('point_in_time',
        datetime.datetime.utcnow().strftime('%Y-%m-%d'))
    delay = config.pop('delay', 30)
    window = config.pop('window': 30)
    range_ = config.pop('range_': 90)

    df_train = D.get(table=input_table,
                     cols=cols,
                     time_col=time_col,
                     time=[point_in_time, delay, window])

    df_labels = D.get(table=target_table,
                      cols=cols,
                      time_col=time_col,
                      time=[point_in_time, delay, window])

    df_predict = D.get(table=input_table,
                       cols=target_cols,
                       time_col=time_col,
                       time=[point_in_time, 0, range_])

    M = ModelFactory.get(project, model, **config)
    M.train(df_train, df_labels)
    df_out = M.predict(df_predict)
    df_out['exec_id'] = exec_id

    S.set('run_output')
    D.put(S, df_out)
```

Code Block 6.20: Example of a runtime script.

```
1   import datetime, argparse
2   import SchemaFactory, DataBaseFactory, ModelFactory
3
4   def main(project, model, mode, db_src, exec_id, **config):
5       # ...
6       # main script here....
7
8   if __name__ == '__main__':
9       parser = argparse.ArgumentParser(description='Inputs for
            Project.')
10      parser.add_argument('--project', type=str, required=True)
11      parser.add_argument('--model', type=str, required=True)
12      parser.add_argument('--mode', type=str, default='TEST')
13      parser.add_argument('--db_src', type=str, default='hive')
14      parser.add_argument('--exec_id', type=int, required=True)
15      # .... additional model configuration parameters
16
17      args = parser.parse_args()
18      main(**args)
```

Code Block 6.21: Example of a runtime script (continued)

model. Additionally, by defining an abstract Model class, as shown in Code Block 6.22, we can ensure that whatever model we deploy will be compatible with our main script, as long as it is subclassed from Model with an implementation of the train and predict methods.

Moreover, by leveraging the principle of polymorphism, we construct our three classes using factory methods, so that our code is agnostic as to which particular subclass of each we are using, as long as that subclass follows the API laid out by its abstract base class. In particular, note that a single script can be leveraged across multiple projects, and many models of the same kind within a project. The model, itself, is simply passed as a parameter into the script, and the factory method delivers the appropriate subclass corresponding to that model. Moreover, we parameterized the point_in_time, so that the same runtime script can be reused in the backtesting framework, as discussed in Section 6.1.6. Another unexpected advantage occurred for one data science team, when their organization migrated to Hive, as we see in our next example.

Example 6.14. A large organization is planning on migrating all its data from Vertica to Hive. One data science team in the organization consists of a number of data scientists who operate mostly independently, writing their own scripts and methods for accessing the database. Moreover, many projects that have been finished have the database queries and connections hard-coded into dusty scripts that are scattered about. The team spends

```
1   from abc import ABC, abstractmethod
2   class Model:
3
4       def __init__(self, **kwargs):
5           # override to provide for default parameter values
6           self.params = kwargs
7
8       @abstractmethod
9       def train(self,
10              df: DataFrame,
11              y: np.array,
12              weights: np.array=None):
13          return
14
15      @abstractmethod
16      def predict(self,
17              df: DataFrame):
18          return
```

Code Block 6.22: Abstract Model class

several months determining all of their production dependencies and over-writing the code in order to point to the new data source. They also have to determine how, exactly, they are going to switch over to Hive, given that the rollout occurs over time and not all of the data is available yet, though they are still constrained to hit the deprecation timeline. The team loses an entire quarter of productivity handling the migration.

Another team, one which embraced the principles of object-oriented design from the ground up, has a much simpler task ahead of it. First, it must write a new subclass from following their `AbstractDataBase` API, replacing the functionality of `DataBase_Vertica` with `DataBase_Hive`. In particular, they must connect to a new data source and update their queries, as there are slight syntactical differences between the two languages. Next, they add one line to their `DataBaseFactory` lookup map:

$$'hive' : DataBaseHive.$$

Finally, they update a single config file that stores `db_src` for all production jobs, changing it to point from `vertica` to `hive`. And Voila! With a few simple changes, the team has migrated all jobs across all projects to the new database. Moreover, they only required help from a single member of their team to write the new database subclass. By using the principles of inheritance, encapsulation, abstraction, and polymorphism, the actual runtime scripts don't need to be changed at all, as they are agnostic with regard to the actual database engine that delivers its data.   ▷

In addition to the migration problem of Example 6.14, there are two other benefits of generating a database class from a factory method. The first is that it might not be possible to connect to Hive from local laptops. Instead, the team might connect to Presto. The factory method allows them to continue to run their jobs locally, by simply passing `db_src='presto'` instead of `db_src='hive'`. The second advantage is when writing unit tests for the code. Sample data can be stored in the form of `.csv` files in a data folder within the project. A new database subclass, `DataBase_Csv`, can then be created to read from the sample datasets in the local folder, requiring no connection to any database. The testing scripts can then pass `db_src='csv'` into the runtime script, so that unit tests will always have access to sample data, without wasting time creating database connections or managing the different connections required based on whether the tests are run locally or on a virtual workstation.

The final advantage of an object-oriented design is that it lubricates collaboration within teams. Since everything follows the same API, different members can easily spin up different models, or try different approaches, in a seamless fashion. One can easily spin up a new model subclass, add the new subclass to the model factory, and then deploy a backtest to see how well the new model would have fared compared to the current production model. Moreover, the modular, a la carte nature of the object-oriented design allows anyone on the team to run any model from anyone else. Such a system is therefore more manageable and maintainable as individual contributors on the team migrate switch roles or migrate to other teams.

Problems

6.1. Construct an counterexample to the claim that $R^2 \geq 0$; i.e., construct a data set and a predictive model with a negative coefficient of determination.

6.2. Show that the softmax transformation defined in Equation (6.12) has the properties $f_i(X) \in (0, 1)$ and $\sum_{i=1}^{m} f_i(X) = 1$.

6.3. Prove Proposition 6.2.

6.4. Determine which of the following discrete variables should be ordinal versus categorical:

1. Age range: $\{18–24, 25–34, 35–44, 45–54, 55\text{-}64, 65+\}$;
2. Gender: {male, female};
3. Year in college: {freshman, sophomore, junior, senior};
4. Brand: {Nike, Adidas, Asics};
5. Compass heading {north, south, east, west};
6. Blood type {A, B, AB, O}.

6.5. Prove that every label encoder is invertible.

6.6. Write your own transformer class in Python that handles one-hot encoding of a categorical variable.

6.7. Write an abstract class for each of the three classes required for the runtime script Code Block 6.20: `AbstractSchema`, `AbstractDataBase`, `AbstractModel`. *Hint*: The SQL syntax for `SELECT`, `CREATE TABLE`, and `INSERT` might vary from database to database, so that `AbstractDataBase` should house the abstract methods to generate the given SQL strings.

Part III

Machine Learning

11

Hello, Neighbor

11.1 Nearest Neighbors

The K-nearest neighbors (KNN) is a simple nonlinear approach to the classification and regression problems that involves making predictions based on the behavior of the "nearest" members in the training set.

11.1.1 Measuring Distance

In this section, we introduce the concept of a metric space, discuss several commonly used distance metrics, and conclude with a brief discussion on the limitations of distance methods.

Metric Spaces

In order to proceed, we first need to establish a notion of "distance." This is achieved mathematically through the following definition.

Definition 11.1. *A* metric space *is a pair* $(\mathcal{M}, d)$, *where* $\mathcal{M}$ *is a set and* $d : \mathcal{M} \times \mathcal{M} \rightarrow \mathbb{R}_*$ *is a nonnegative function, known as a* metric, *that satisfies the properties*

1. **(symmetry)** $d(x, y) = d(y, x)$,
2. **(identifiability)** $d(x, y) = 0$ *if and only if* $x = y$,
3. **(triangle inequality)** $d(x, y) + d(y, z) \geq d(x, z)$,

for all $x, y, z \in \mathcal{M}$. *A metric space is* complete *if every* Cauchy sequence[1] *converges to a point in* $\mathcal{M}$.

---

[1] A sequence $\{x_i\}$ in a metric space M is a *Cauchy sequence* if for every $\epsilon > 0$ there exists a n_ϵ such that $d(x_i, x_j) < \epsilon$ for all $i, j > n_\epsilon$. Counterexample: the open interval $(0, 1)$ is not complete.

Common Distance Metrics

There have been a multitudinous plethora of metrics that have been dis-
covered. (The author has even published a novel metric for orbits in as-
trodynamics; see Maruskin [2010].) Another fun example is the French-rail
metric, which is constructed based on the observation that all rail lines in
France lead to Paris; see Exercise 11.1. In this paragraph, we will discuss
several key metrics commonly used in data science.

Definition 11.2. *The* Minkowski distance *of order p in $\mathbb{R}^k$ is the metric*

$$d_p(x, y) = ||x - y||_p = \left(\sum_{j=1}^{k} |x_j - y_j|^p \right)^{1/p}, \qquad (11.1)$$

for $x, y \in \mathbb{R}^k$, where $|| \cdot ||_p$ is the ℓ_p norm, defined in Definition 7.4.

In particular, for $p = 1$, we have the Manhattan distance; *for $p = 2$
we have the* Euclidean distance, *and for $p = \infty$, we have the* Chebyshev
distance, *given by the formulas*

$$d_1(x, y) = \sum_{j=1}^{k} |x_j - y_j|, \qquad (11.2)$$

$$d_2(x, y) = \left(\sum_{j=1}^{k} |x_j - y_j|^2 \right)^{1/2}, \qquad (11.3)$$

$$d_\infty(x, y) = \max_{j=1,\dots,k} |x_j - y_j|, \qquad (11.4)$$

respectively.

The Chebyshev distance has an interesting interpretation in terms of
the movement of the king on a chessboard; see Exercise 11.2.

Though not strictly a Minkowski metric, if we define $0^0 = 0$, we can
formally express Equation (11.1) for the case $p = 0$ as

$$d_0(x, y) = \sum_{j=1}^{k} \mathbb{I}[x_j = y_j]. \qquad (11.5)$$

When $x, y \in \mathbb{B}^k$ are binary vectors, this metric is known as the *Hamming
distance*, which counts the number of bits in vector x that differ from the
corresponding bits in vector y. The Hamming distance is important when
measuring distance in a one-hot encoded categorical feature space.

The Curse of Dimensionality

The *curse of dimensionality* is a phenomenon that occurs in certain learning methods that describes a severe loss in performance with increased dimensionality due to the increasing vastness of the larger space.

To understand what we mean by the *increase in vastness*, consider an n-dimensional sphere of radius r in $\mathbb{R}^n$, whose volume is given by

$$V_n(r) = \frac{\pi^{n/2} r^n}{\Gamma(1 + n/2)}.$$

The ratio of this volume to the volume $2^n r^n$ of the inscribed hypercube of side-length 2 decays to zero as $n \to \infty$. In particular, the volume $V_n(1)$ of the unit sphere is plotted for $n = 1, \ldots, 20$ in Figure 11.1. Note that the volume of the unit sphere decays to zero with increasing dimensionality. This implies that the fraction of instances with a distance less than unity becomes vanishingly small as the dimensionality increases.

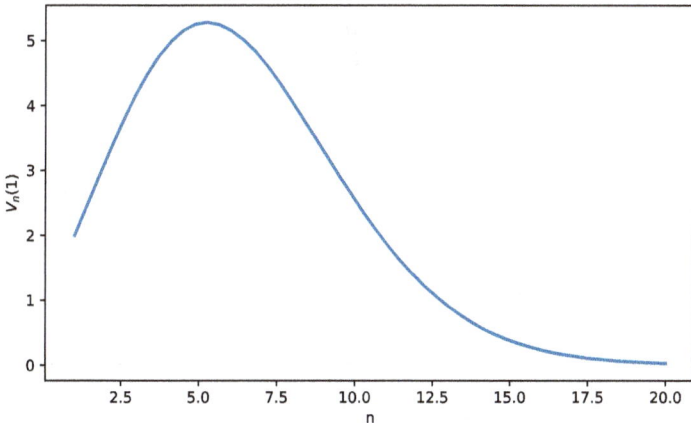

Fig. 11.1: The volume $V_n(1)$ of a unit sphere as a function of dimension n.

Moreover, the distance between the origin and the corner $(1, \ldots, 1)$ of the unit hypercube is $\sqrt{n}$, which increases without bound as $n \to \infty$. This implies that most of the space is "far away" from the origin, even within the unit cube. Thus, any concept of *nearness* decays commensurately with increased dimensionality. For this reason, the efficacy of nearest-neighbor algorithms wanes as the dimensionality increases, and such algorithms are ill suited for high-dimensional problems.

11.1.2 Finding the Nearest Neighbors

Now that we have a distance function in hand, we can easily compute the k nearest neighbors to a given point x_0 from a training data set $\mathcal{D}$. The details are shown in Algorithm 11.1.

Naturally, in order to locate the nearest neighbors, we need to first compute the distance between the test point x_0 and each point in the data set $\mathcal{D}$, which is an $O(n)$ operation, as we must iterate through the given dataset.

Next, we must locate the smallest values in our computed set of distances. Now, if we sort the entire dataset, this requires at best $O(n \log n)$. However, since we only need to find the k smallest values, and we do not need to sort the entire dataset, we can achieve this in $O(nk)^2$. We represent this method as $\texttt{argmin}(d, k)$, as shown on line 6. In Python, this can be achieved using $\texttt{np.argpartition(d, k)}$, which will return an array of indices, such that the first k elements of said array are guaranteed to correspond to the indices of the k smallest values in the array d of distances.

11.1.3 Supervised Learning with Nearest-Neighbors

Now that we have defined distance in our feature space, we are ready to dive into the k-nearest neighbors (KNN) algorithm for classification and regression. At a high level, the algorithm proceeds in two steps: locate your k nearest neighbors and then *do what your neighbors do*. For classification, this involves letting your nearest neighbors vote, using their own labels, on you predicted class. For regression, this involves averaging the target variable across your nearest neighbors.

Algorithm 11.1: findNeighbors($\mathcal{D}$, x_0, k) – algorithm for finding k nearest neighbors.

Input: data $\mathcal{D} = \{(x_i, y_i)\}_{i=1}^n$;
  a feature vector x_0;
  the number k of nearest neighbors;
  a distance metric $\texttt{dist}()$
Output: The set $(\iota_1, \ldots, \iota_k)$ of indices for the k nearest neighbors.
1 Set $n = |\mathcal{D}|$
2 Set $d = \texttt{Array}()$
3 **for** $i = 0, \ldots, n-1$ **do**
4 | Set $d[i] = \texttt{dist}(x_i, x_0)$
5 **end**
6 Set $(\iota_1, \ldots, \iota_k) = \texttt{argmin}(d, k)$
7 **return** $(\iota_1, \ldots, \iota_k)$

---

[2] For example, by using the *bubble-sort algorithm*; see Exercise 11.3 and Lambert [2014].

We begin by laying out the nearest neighbors algorithms for classification and regression. We then discuss various aspects of the two algorithms before concluding with several illustrative examples.

The Nearest Neighbors Classifier

The k-nearest neighbors (KNN) classifier involves seeking the k training instances closest to a given test point, and then allowing those nearest neighbors to vote on the predicted class for the test point. Once we have computed the target values of the k nearest neighbors, we let each of the nearest neighbors "vote" on the prediction for the test-point x_0. To do this, we calculate the ratio of neighbors that belong to each class $\mathcal{C}$, as shown in line 4 of Algorithm 11.2. Finally, we return the argmax of p; i.e., the most popular class of the nearest neighbors. We should take care to break ties with random chance.

The Nearest Neighbors Regressor

The algorithm for the KNN classifier can be easily modified to accommodate regression problems. Essentially, instead of polling the nearest neighbors to determine the most popular class, we instead return the average values of the neighborhood, as shown in Algorithm 11.3.

Neighborhood Watch

For both classification and regression, the nearest neighbors algorithms present with a few limitations that one should keep in mind.

Algorithm 11.2: knnClassifier($\mathcal{D}$, x_0, k) – k-nearest neighbors algorithm for classification.

Input: data $\mathcal{D}$;
a feature vector x_0;
the number k of nearest neighbors;
the set $\mathcal{C}$ of target classes
Output: A predicted class $\hat{y}$.
1 Set $(\iota_1, \ldots, \iota_k) = \texttt{findNeighbors}(\mathcal{D}, x_0, k)$
2 Set $p = \texttt{Dict}()$
3 **for** c in $\mathcal{C}$ **do**
4   Set $p[c] = \dfrac{1}{k} \sum_{i=1}^{k} \mathbb{I}[y_{\iota_i} = c]$
5 **end**
6 **return** $\texttt{argmax}(p)$

Curse of Dimensionality

Nearest neighbors is a prime example of the curse of dimensionality. Since the algorithm heavily relies on the distance metric and the presumption that one can locate a collection of *nearest neighbors* that are indicative of a given test point's label, it therefore suffers drastically from the vastness encountered when one uses more than a handful of features. We saw how, in high dimensional spaces, the volume of the unit ball quickly decays to zero, and the distance to the edge of the unit cube grows without bound. Thus, when one uses too many dimensions, one quickly discovers that all neighbors become rather far away. This explosion of distance and vastness breaks the correlation between the given test point and its neighbors.

Feature Scaling

Given a problem with a low-dimensional feature space, we may still encounter issues if the continuous features span different scales from one another. For example, suppose we are using nearest neighbors to predict whether a house will sell within the first seven days it goes on market, based on the number of schools and the average listing price in the literal neighborhood. There is a full fivefold gap in the orders of magnitude between the number of schools $O(10^0)$ and the average listing price $O(10^5)$. This has the practical effect of negating the predictive power of the former variable.

When implementing nearest neighbor algorithms, it is therefore critical to center and scale the continuous predictors, so that they may be properly compared using the Euclidean metric.

Selecting k

In the KNN algorithm, the parameter k should be viewed as a tunable hyperparameter for the model. If k is too small, the model will have low bias and high variance. At the extreme end, if $k = 1$ and the training data is noisy, the resulting predictions will simply follow the noise in the training

---

Algorithm 11.3: knnRegressor($\mathcal{D}$, x_0, k) – k-nearest neighbors algorithm for regression.

Input: data $\mathcal{D}$;
    a feature vector x_0;
    the number k of nearest neighbors;
    the set $\mathcal{C}$ of target classes
Output: A predicted value $\hat{y}$.
1 Set $(\iota_1, \ldots, \iota_k) = \text{findNeighbors}(\mathcal{D}, x_0, k)$
2 return $\dfrac{1}{k} \sum_{i=1}^{k} y_{\iota_i}$

---

set. If k is too large, on the other hand, the model will have high bias and low variance. At the extreme end on this side of the spectrum, if $k = n$, the model will simply predict the overall average without regard to location in feature space.

To find the optimal value of k, we can therefore run the model over a range of possible values of k, and select the value that minimizes the generalization error (Definition 6.21). We can therefore settle upon an appropriate value of k, for a given problem, through the usual process of model tuning.

Large Datasets

Training a nearest neighbor model is tantamount to memorizing the training dataset. To apply a prediction, we have to iterate through the entire training set, calculating the distance between each training instance and the test point. Thus, the prediction step is at least $O(n)$. Moreover, the entire training set must be retained in memory. Thus, the algorithm becomes burdensome for large datasets.

To remedy this, it is natural to seek to replace the training dataset with a less memory-intensive representation of the same. For example, one might store the data in a multidimensional binary search tree, due to Bentley [1975], for which retrieval of the k nearest neighbors can be achieved in $O(\log n)$.

Distance Weighting (A Variation)

Finally, we mention a natural variation of the k-nearest neighbors algorithm, which helps reduce the frequency of ties. Given the indices $(\iota_1, \ldots, \iota_k)$ of the k-nearest neighbors, we may calculate the weights

$$w_i = \frac{1}{d(x_0, x_{\iota_i})},$$

for $i = 1, \ldots, k$. Using these inverse-distance weights, we may then modify Algorithms 11.2 and 11.3 by the weighted averages

$$p[c] = \frac{1}{w} \sum_{i=1}^{k} w_i \mathbb{I}[y_{\iota_i} = c] \quad \text{and} \quad \hat{y} = \frac{1}{w} \sum_{i=1}^{k} w_i y_i,$$

respectively, where $w = \sum_{i=1}^{k} w_i$. In particular, we could even let $k = n$, and still retain unique predictions for various regions of the feature space.

Python Implementation

A k-nearest neighbors class, that is valid for both classification and regression, is given in Code Block 11.1. This is an implementation of the most basic algorithm, and suffers from the deficits mentioned in the previous paragraph. See also the `KNeighborsClassifier` and `KNeighborsRegressor` classes from the `sklearn.neighbors` module.

```
1   class KNN(Model):
2
3       def __init__(self, **kwargs):
4           self.k = kwargs['k']
5           self.type = kwargs['type']
6           assert self.type in ('classification', 'regression')
7
8       def train(self, df: DataFrame, y: np.array, weights:
                np.array=None):
9           X = df.values if isinstance(df, DataFrame) else df
10          self.X = X
11          self.y = y
12
13      def predict(self, df: DataFrame):
14          X = df.values if isinstance(df, DataFrame) else df
15          y = np.zeros(len(X))
16          for i in range(len(X)):
17              d = [np.linalg.norm(self.X[j, :] - X[i,:]) for j in
                    range(len(self.y))]
18              indices = np.argpartition(d, self.k)[:self.k]
19              if self.type == 'classification':
20                  y[i] = Counter(self.y[indices]).most_common(1)[0][0]
21              else:
22                  y[i] = np.mean(self.y[indices])
23          return y
```

Code Block 11.1: k-nearest neighbors class.

Examples

Example 11.1. We can simulate a two-dimensional binary classification problem with decision boundary $x_1 = \sin(x_0)$, such that the probability a given instance belongs to the class 1 is given by

$$p(x_0, x_1) = \frac{1}{1 + e^{-\alpha(x_1 - \sin(x_0))}}.$$

A sample of 200 training instances generated in such a manner, with $\alpha = 2$, are shown in Figure 11.2. The code used to simulate this training set is given in Code Block 11.2.

Next, we build a meshgrid over the (x_0, x_1) plane, of increment size 0.1, and calculate the nearest neighbors predictions for each point on our grid. The code is given in Code Block 11.2. The nearest neighbor predictions are shown visually in Figures 11.3 and 11.4 for $k = 1, 3, 5, 10$, and 100.

Notice the extremes: the case $k = 1$ (Figure 11.3 (top)) overfits the noise in the training set, whereas the case $k = 100$ (Figure 11.4 (bottom)) has degenerated to a nearly linear diagonal decision boundary.   ▷

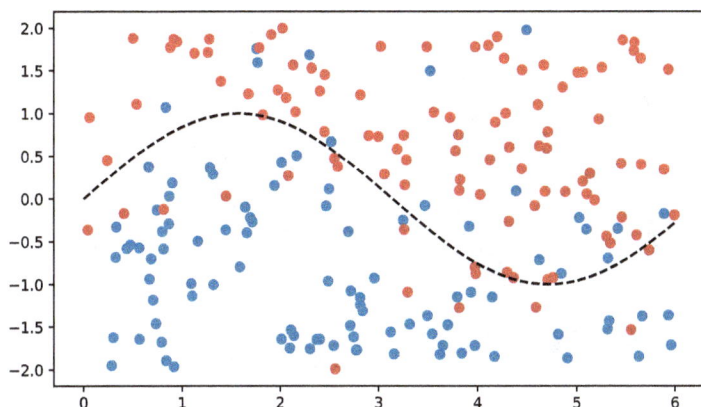

Fig. 11.2: A simulated classification problem with decision boundary $x_1 = \sin(x_0)$ and $n = 200$ random training instances.

```
n = 200
x = np.random.uniform(0, 6, size=n)
y = np.random.uniform(-2, 2, size=n)
X = np.array([x, y]).T
p = (1 + np.exp(-alpha*(y - np.sin(x) )))**(-1)
y = (np.random.random(size=n) < p).astype(int)

X_grid, Y_grid = np.meshgrid(np.linspace(0, 6, 61), np.linspace(-2,
    2, 41))
XX = np.array([X_grid.reshape((2501)), Y_grid.reshape((2501))]).T

K = KNN(k=5, type='classification')
K.train(X, y)
y_hat = K.predict(XX).reshape((41,61))

cm = colors.LinearSegmentedColormap.from_list('mylist', ['#1f77b4',
    '#d62728'], N=2)
plt.pcolormesh(X_grid, Y_grid, y_hat, cmap=cm)
plt.plot(t, np.sin(t), 'w')
plt.title(f"k={k}")
```

Code Block 11.2: Code to simulate training instances and calculate grid-wise nearest neighbor predictions.

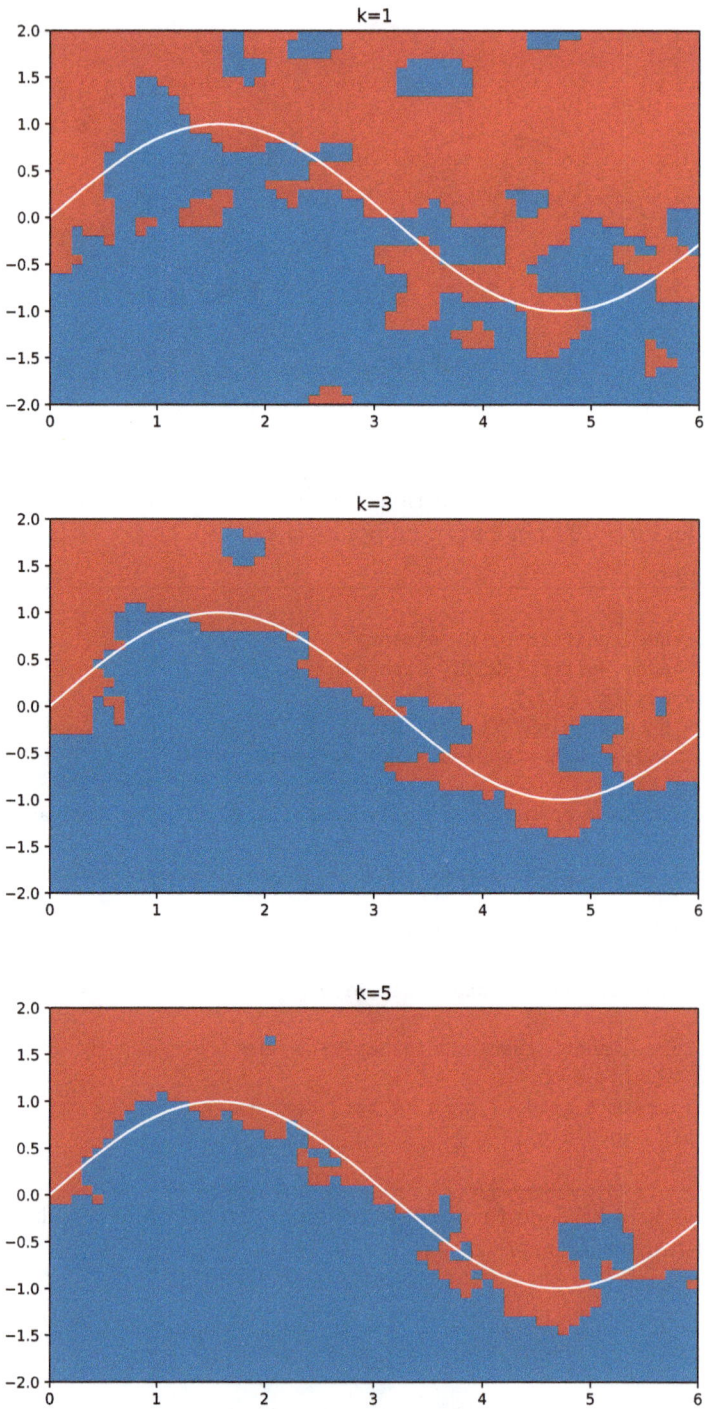

Fig. 11.3: Nearest Neighbors Predictions for $k = 1, 3, 5$.

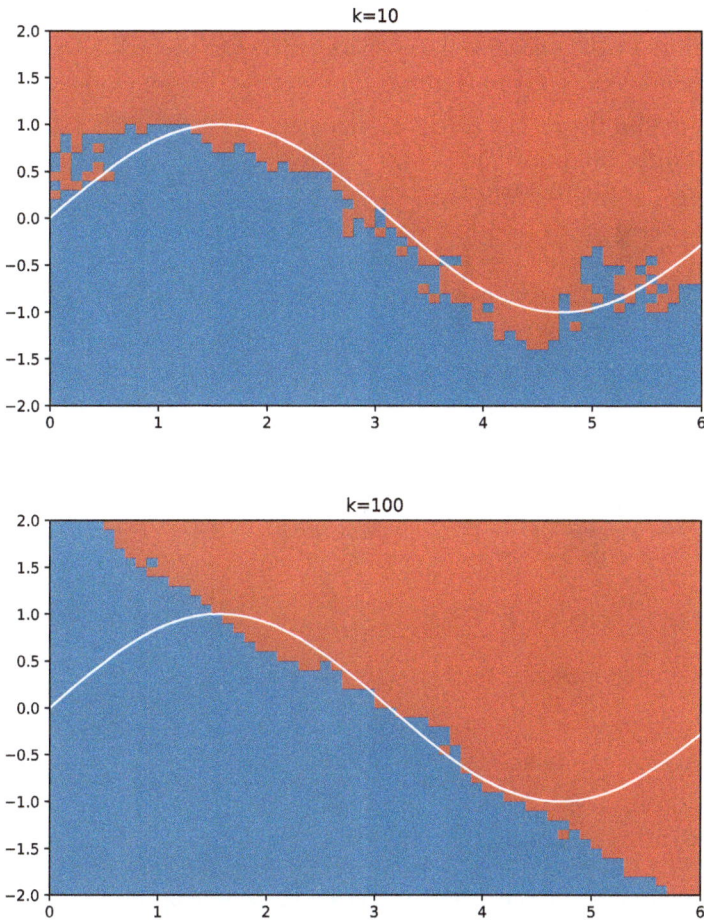

Fig. 11.4: Nearest Neighbors Predictions for $k = 10, 100$.

11.1.4 Inner Product Spaces

Note that metric spaces are more general than Euclidean space, and, indeed, need not even represent a vector space. It is therefore useful to briefly recall the following definitions. For additional details, see DiBenedetto [2002] or Folland [1999].

Normed Linear Space

Definition 11.3. *A* normed linear space *is a vector space* $\mathbb{X}$ *over a field* $\mathcal{F}$[3] *with a function* $|| \cdot || : \mathbb{X} \to \mathbb{R}^*$, *called a* norm, *that satisfies the properties*

1. **(identifiability)** $||x|| = 0$ *if and only if* $x = 0$,
2. **(triangle inequality)** $||x + y|| \leq ||x|| + ||y||$,
3. **(scalar multiplication)** $||\lambda x|| = |\lambda| \, ||x||$,

for all $x, y \in \mathbb{X}$ *and* $\lambda \in \mathcal{F}$.

As it turns out, every normed linear space is a metric space, though the contrapositive need not be true.

Proposition 11.1. *Given a normed linear space* $\mathbb{X}$, *the function* $d : \mathbb{X} \times \mathbb{X} \to \mathbb{R}^*$ *defined by the relation*

$$d(x, y) = ||x - y|| \tag{11.6}$$

satisfies the axioms of a metric, so that the pair $(\mathbb{X}, d)$ *constitutes a metric space.*

Proof. See Exercise 11.4. □

For completeness, we further state the following.

Definition 11.4. *A* Banach space *is a normed linear space that is complete with respect to the induced metric Equation* (11.6).

Inner Product Spaces

Euclidean spaces still have more structure than pure normed linear spaces.

Definition 11.5. *An* inner product space *is a vector space* $\mathbb{X}$ *over* $\mathbb{C}$ *(or* $\mathbb{R}$*) with an* inner product $\langle \cdot, \cdot \rangle : \mathbb{X} \times \mathbb{X} \to \mathbb{C}$ *that satisfies the axioms*

1. **(symmetry)** $\langle x, y \rangle = \overline{\langle y, x \rangle}$,
2. **(linearity)** $\langle \lambda x + y, z \rangle = \lambda \langle x, z \rangle + \langle y, z \rangle$,
3. **(positive definite)** $\langle x, x \rangle \geq 0$,
4. **(identifiability)** $\langle x, x \rangle = 0$ *if and only if* $x = 0$,
5. **(triangle inequality)** $d(x, y) + d(y, z) \geq d(x, z)$,

for all $x, y, z \in \mathbb{X}$ *and* $\lambda \in \mathbb{C}$.

As it turns out, every inner product space is also a normed linear space, and therefore a metric space.

---

[3] typically, the field is $\mathbb{R}$ or $\mathbb{C}$, in which event, we call $\mathbb{X}$ a *real* or *complex vector space*, respectively.

Proposition 11.2. *Given an inner product space* $\mathbb{X}$, *the function* $|| \cdot || :$ $\mathbb{X} \to \mathbb{R}^*$ *defined by the relation*

$$||x||^2 = \langle x, x \rangle \tag{11.7}$$

satisfies the axioms of a norm, so that the pair $(\mathbb{X}, ||\cdot||)$ *constitutes a normed linear space.*

Proof. See Exercise 11.5. □

Definition 11.6. *A* Hilbert space *is an inner product space that is complete with respect to the metric induced via Equations (11.7) and (11.6).*

11.2 K-Means Clustering

In this section, we discuss our first unsupervised learning algorithms, which seek to uncover hidden structures in unlabelled data sets. In particular, we seek to identify structures known as *clusters*. For additional details, see standard texts, such as Hastie, *et al.* [2009] or James, *et al.* [2013]. For details of implementation in Python with sci-kit learn, see Gèron [2019].

11.2.1 Clustering

We begin by introducing notation that allows us to define and describe clusters and a cluster assignment, given a set of unlabelled data $\mathcal{X}$.

Definition 11.7. *Given a set of unlabelled data* $\mathcal{X}$ *of size* $n = |\mathcal{X}|$ *and a number* $c \in \mathbb{Z}^*$, *a* cluster assignment *is an many-to-one mapping (i.e., an encoding)* $\alpha : \iota(\mathcal{X}) \to \{1, \ldots, c\}$, *where* $\iota(\mathcal{X})$ *is the indexing set for* $\mathcal{X}$, *that assigns each datum* $x_i \in \mathcal{X}$ *to the unique cluster with label* $\alpha(i)$.

Given a cluster assignment α, *the* kth *cluster is the subset* $\mathcal{C}_k \subset \mathcal{X}$ *defined by*

$$\mathcal{C}_k = \{x_i : \alpha(i) = k\}, \tag{11.8}$$

for $k = 1, \ldots, c$.

The cluster indicator *is the function* $\delta : \iota(\mathcal{X}) \times \{1, \ldots, c\} \to \mathbb{B}$ *defined by*

$$\delta_{ik} = \mathbb{I}[\alpha(i) = k], \tag{11.9}$$

for $i \in \iota(\mathcal{X})$ *and* $k = 1, \ldots, c$.

The set $\mathcal{C} = \{\mathcal{C}_1, \ldots, \mathcal{C}_c\}$ therefore constitutes a *partition* of the set $\mathcal{X}$, since the clusters are pairwise disjoint, and their union covers the set $\mathcal{X}$. The goal of the *clustering problem* is to *learn* a cluster assignment that minimizes some property, typically the within-group variance.

11.2.2 Partitioning Theorem for Inner Product Spaces

Before discussing our first clustering method, we first consider a general-ization of the partitioning theorem for inner product spaces. We begin by defining the following sums of squares for a normed linear space.

Definition 11.8. *Let* $(\mathbb{X}, d)$ *be a normed linear space and* $\mathcal{X} \subset \mathbb{X}$. *For any cluster assignment* α, *we define the* between-group sum of squares *(SSB)*, *the* within group sum of squares *(SSW)*, *and the* total sum of squares *(SST) are defined by*

$$\text{SSB} = \sum_{k=1}^{c} r_k \left|\left| \bar{x}_k - \bar{\bar{x}} \right|\right|^2, \tag{11.10}$$

$$\text{SSW} = \sum_{i=1}^{n} \left|\left| x_i - \bar{x}_{\alpha(i)} \right|\right|^2, \tag{11.11}$$

$$\text{SST} = \sum_{i=1}^{n} \left|\left| x_i - \bar{\bar{x}} \right|\right|^2, \tag{11.12}$$

where we have defined

$$r_k = \sum_{i=1}^{n} \delta_{ik}, \qquad \bar{x}_k = \frac{1}{r_k} \sum_{i=1}^{n} x_i \delta_{ik}, \qquad \bar{\bar{x}} = \frac{1}{n} \sum_{i=1}^{n} x_i.$$

We require a normed linear space, and not simply a metric space, as the vector-space structure is required to define the means in relation to the vector sums. If our space also has an inner product, we have the following partitioning theorem, analogous to Theorem 3.11.

Theorem 11.1 (Partitioning Theorem for Inner Product Spaces). *Let* $\mathbb{X}$ *be an inner-product space,* $\mathcal{X} \subset \mathbb{X}$, *and* α *a cluster assignment. Then the following identity holds*

$$\sum_{i=1}^{n} \left|\left| x_i - \bar{\bar{x}} \right|\right|^2 = \sum_{i=1}^{n} \left|\left| x_i - \bar{x}_{\alpha(i)} \right|\right|^2 + \sum_{k=1}^{c} r_k \left|\left| \bar{x}_k - \bar{\bar{x}} \right|\right|^2, \tag{11.13}$$

where $|| \cdot ||$ *is the norm induced by the inner product.*

Note that relative to the natural metric on $\mathbb{X}$, Equation (11.13) is equiv-alent to the class sum-of-squares decomposition

$$\text{SST} = \text{SSB} + \text{SSW}.$$

Proof. To proceed, let us reindex the set $\mathbb{X}$ using the notation x_{kj} to repre-sent the jth elements of cluster $\mathcal{C}_k$, for $k = 1, \ldots, c$ and $j = 1, \ldots, r_k$. Next, we can express

$$\sum_{i=1}^{n} ||x_i - \bar{\bar{x}}||^2 = \sum_{k=1}^{c} \sum_{j=1}^{r_k} ||x_{kj} - \bar{\bar{x}}||^2$$

$$= \sum_{k=1}^{c} \sum_{j=1}^{r_k} ||(x_{kj} - \bar{x}_k) + (\bar{x}_k - \bar{\bar{x}})||^2$$

$$= \sum_{k=1}^{c} \sum_{j=1}^{r_k} \left\{ ||x_{kj} - \bar{x}_k||^2 + 2 \left\langle x_{kj} - \bar{x}_k, \bar{x}_k - \bar{\bar{x}} \right\rangle + ||\bar{x}_k - \bar{\bar{x}}||^2 \right\},$$

where the final step holds for real inner-product spaces. Now, the middle term sums to zero, since

$$\sum_{k=1}^{c} \sum_{j=1}^{r_k} \left\langle x_{kj} - \bar{x}_k, \bar{x}_k - \bar{\bar{x}} \right\rangle = \sum_{k=1}^{c} \left\langle \sum_{j=1}^{r_k} (x_{kj} - \bar{x}_k), \bar{x}_k - \bar{\bar{x}} \right\rangle$$

$$= \sum_{k=1}^{c} \left\langle r_k \bar{x}_k - r_k \bar{x}_k, \bar{x}_k - \bar{\bar{x}} \right\rangle = 0.$$

For complex inner-product spaces, the term $2 \left\langle x_{kj} - \bar{x}_k, \bar{x}_k - \bar{\bar{x}} \right\rangle$ is replaced with this inner product plus its complex conjugate; since the term vanishes, however, the result still holds. Identifying

$$\sum_{i=1}^{n} ||x_i - \bar{x}_{\alpha(i)}||^2 = \sum_{k=1}^{c} \sum_{j=1}^{r_k} ||x_{kj} - \bar{x}_k||^2$$

$$\sum_{k=1}^{c} r_k ||\bar{x}_k - \bar{\bar{x}}||^2 = \sum_{k=1}^{c} \sum_{j=1}^{r_k} ||\bar{x}_k - \bar{\bar{x}}||^2$$

completes our proof.   □

11.2.3 K-Means Clustering Algorithm

The goal of the K-means clustering algorithm is to find a cluster assignment that minimizes the within-group sum of squares Equation (11.11). The idea is define the assignment α relative to a set of c vectors $m_1, \ldots, m_c \in \mathbb{X}$, which may be thought of as *centers of gravity* for our clusters. We then assign each datum x_i to the cluster k that has the closest center m_k. The full algorithm is shown in Algorithm 11.4.

While Algorithm 11.4 is guaranteed to converge, it is not guaranteed to converge to the global minimum. It is therefore customary to run the algorithm multiple times, and choose the output with the minimum value of ssw. The Python code for the K-means algorithm is shown in Code Block 11.3.

In order to test our method, we can simulated data using a Gaussian mixture model with random centers, as shown in Code Block 11.4. Once we have simulated data, we can run our K-means algorithm and plot the results, which are shown in Figure 11.5.

Algorithm 11.4: kMeans($\mathcal{X}$, c) – algorithm for finding c clusters over a normed linear space.

Input: data $\mathcal{X} = \{x_i\}_{i=1}^n$ belonging to an inner product space $\mathbb{X}$; the number c of clusters

Output: The set $\{m_1, \ldots, m_k\}$ of cluster means.

1 Let $(i_1, \ldots, i_n)$ be a random shuffle of $(1, \ldots, n)$
2 Set $m_k = x_{i_k}$ for $k = 1, \ldots, c$
3 Set $\Delta = 1$
4 **while** $\Delta > 0$ **do**
5   Calculate the assignment $\alpha(i) = \underset{k=1,\ldots,c}{\arg\min}\, d(x_i, m_k)$
6   Let $\delta_{ik} = \mathbb{I}[\alpha(i) = k]$ and $r_k = \sum\limits_{i=1}^n \delta_{ik}$
7   **for** $k = 1, \ldots, c$ **do**
8   Set $m'_k = \dfrac{1}{r_k} \sum\limits_{i=1}^n x_i \delta_{ik}$
9   **end**
10   Set $\Delta = \sum\limits_{k=1}^c (m_k - m'_k)^2$
11   Set $m_k = m'_k$, for $k = 1, \ldots, c$
12 **end**
13 **return** $(m_1, \ldots, m_k)$

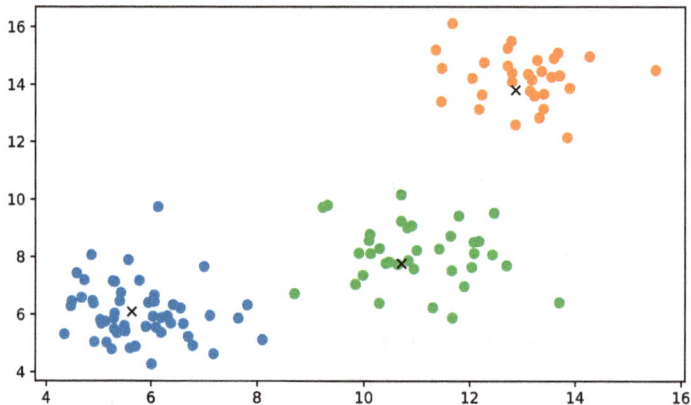

Fig. 11.5: Simulated clusters and final output of K-means algorithm.

```
 1   class KMeans:
 2
 3       def __init__(self, c):
 4           self.c = c
 5
 6       def getAssignment(self, X, kmeans):
 7           n, d = X.shape
 8           c, _ = kmeans.shape
 9           distances = np.zeros((n, c))
10           for k in range(c):
11               m = kmeans[k, :]
12               distances[:, k] = np.linalg.norm(X - m, axis=1)
13           return np.argmin(distances, axis=1)
14
15       def train(self, X):
16           n, d = X.shape
17           indy = np.random.choice(np.arange(n), size=self.c,
                   replace=False)
18           kmeans = X[indy, :]
19           go = True
20           while go:
21               # Compute Current Assignment
22               assignment = self.getAssignment(X, kmeans)
23               rk = dict(Counter(assignment))
24               rk = [rk.get(k, 0) for k in range(self.c)]
25               # Compute New Means
26               kmeans_new = np.zeros(kmeans.shape)
27               for k in range(self.c):
28                   kmeans_new[k, :] = X[assignment==k, :].mean(axis=0)
29               go = (kmeans_new == kmeans).any()
30               kmeans = kmeans_new
31           self.kmeans, self.assignment = kmeans, self.getAssignment(X,
                   kmeans)
32           return self.kmeans, self.assignment
```

Code Block 11.3: *K*-Means clustering algorithm.

11.2.4 Clusters as Features

Clustering can also be used to find structures in long-tail categorical features. A common case is that of *customer country*. There are 195 recognized countries in the world, though data are not distributed evenly across countries. It is common the case to have a collection of larger countries that constitutes the bulk of the user base, and a *long tail* of smaller countries, for which there is insufficient data to draw statistically significant conclusions. We therefore seek to uncover similarities in the long-tail countries, in

```
1  c = 3
2  locs = np.random.randint(5, 20, size=(c, 2))
3  rs = np.random.randint(20, 60, size=c)
4  data = []
5  for i in range(c):
6      data.append(np.random.normal(loc=locs[i,:], size=(rs[i], 2)))
7  X = np.concatenate(data)
8  K = KMeans(c)
9  kmeans, alpha = K.train(X)
10 for k in range(c):
11     plt.plot(X[alpha==k,0], X[alpha==k,1], 'o')
12     plt.plot(kmeans[k,0], kmeans[k,1], 'kx')
```

Code Block 11.4: Simulation of clusters and K-means solution

the hopes that groups of countries might behave similarly, so that we can reduce the effective dimensionality of our learning problem.

One approach is to seek to uncover similarities based on actual data for a given application. For example, one could use observed average marketing indicators (cost-per-click, click-through-rate, etc.) to form clusters.

Another approach is to use enriched features, that can be gathered from third party sources. A fantastic trove of information is the World Bank. One can write calls to the World Bank's REST API; however there is an open-source Python package, wbdata (`wbdata.readthedocs.io`), that one can use more readily. We pull population and GDP per capita in Code Block 11.5, and cluster countries based on the log values. The data are plotted in Figure 11.6.

The resulting clusters are shown in Figure 11.6. We can visualize these results geographically with the **geopandas** package (`geopandas.org`), as shown in Figure 11.7. Note that the distribution over population and GDP per capita is a multivariate lognormal distribution, i.e., data are normally distributed on a log-log plot. Because of this, the resulting clusters are largely arbitrary as there are no natural "breaks" in the data. Due to the sensitivity of the algorithm on the initial (random) K-means, no two results will be the same. Nonetheless, each cluster represents a group of geos that have a similar size and income level.

The World Bank data is not limited to population and GDP; it houses hundreds of country-level indicators in a broad range of categories, like housing, education, debt, gender, poverty, wealth, economics, and so forth. Each of these indicators is also a time-series indicator, so interesting patterns can be uncovered by plotting indicators over time. When combined with internal application-specific data, one can uncover hidden structures and craft relevant segments for the particular business use case.

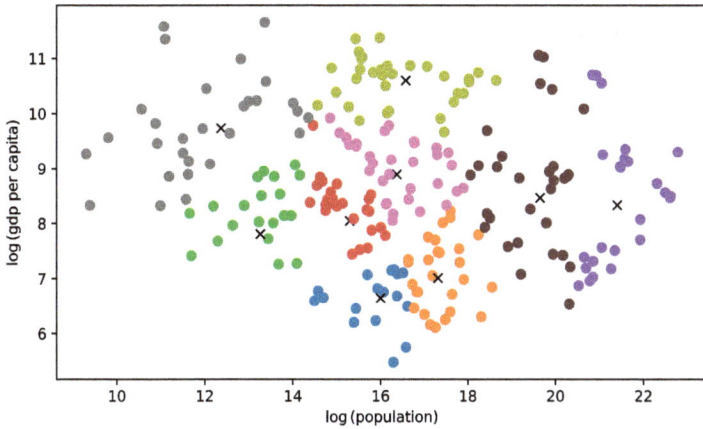

Fig. 11.6: Clustering countries by population and GDP per capita (log scale), using $c = 9$ clusters.

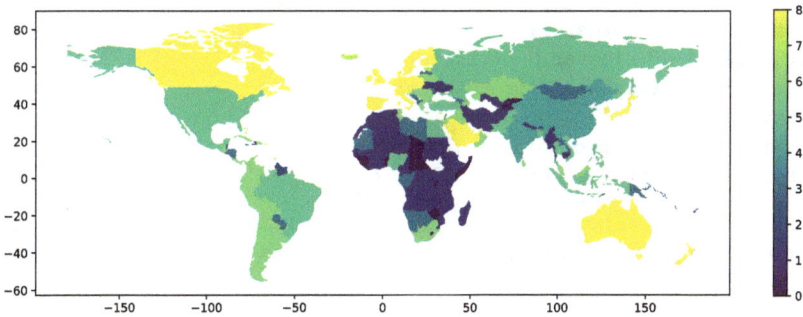

Fig. 11.7: Population-GDP clusters shown on a map.

11.2.5 Scatter

The K-means clustering algorithm relies on the underlying feature set to derive from an inner product space (and hence a vector space), so that the means may be calculated in line 8 and so we may compute the SSW of Equation (11.11). This strongly relies on the condition that we only include numerical features. For categorical features or, more generally, for measures of association based on general similarity rules, it is natural to ask if we can construct a comparable algorithm when our feature set is a metric space, but not a vector space. For such a generalization, we may no longer rely on averages, such as $\bar{x}_k$ or $\bar{\bar{x}}$. For these purposes, we introduce a generalized version of variance known as *scatter*.

```
1   import geopandas as gpd
2   import wbdata
3   dates = datetime.datetime(2020,1,1), datetime.datetime(2021,1,1)
4   gdp_per_capita = wbdata.get_data("NY.GDP.PCAP.CD", data_date=dates)
5   population = wbdata.get_data("SP.POP.TOTL", data_date=dates)
6   df = DataFrame(columns=['geo', 'population', 'gdp_per_capita'])
7   for item in population:
8       df.loc[item['country']['id'], 'geo'] = item['countryiso3code']
9       df.loc[item['country']['id'], 'population'] = item['value']
10  for item in gdp_per_capita:
11      df.loc[item['country']['id'], 'geo'] = item['countryiso3code']
12      df.loc[item['country']['id'], 'gdp_per_capita'] = item['value']
13  df = df[~df.gdp_per_capita.isnull()]
14  df[['population', 'gdp_per_capita']] = np.log(df[['population',
        'gdp_per_capita']].astype(float))
15  X = df[['population', 'gdp_per_capita']].values
16  c = 9
17  K = KMeans(c)
18  kmeans, alpha = K.train(X)
19  df.loc[:, 'cluster'] = alpha
20  order = Series(kmeans[:,1]).sort_values().index.tolist()
21  cluster_order = {order[i]: i for i in range(c)}
22  df.cluster = df.cluster.map(cluster_order)
23  alpha = Series(alpha).map(cluster_order).values
24  countries =
        gpd.read_file(gpd.datasets.get_path("naturalearth_lowres"))
25  countries = countries.merge(df, left_on='iso_a3', right_on='geo')
26  countries.plot("cluster", legend=True)
```

Code Block 11.5: Clustering based on population and GDP per capita.

Definition 11.9. *The* scatter *of a finite subset $\mathcal{X}$ of a metric space (X, d) is defined by the relation*

$$\text{SCAT}(\mathcal{X}) = \frac{1}{2n} \sum_{i=1}^{n} \sum_{j=1}^{n} d(x_i, x_j)^2, \qquad (11.14)$$

where $n = |\mathcal{X}|$.

The within-group scatter *of $\mathcal{X}$ with respect to a cluster assignment $\alpha : \iota(\mathcal{X}) \to (1, \ldots, c)$ is defined as*

$$\text{SCW} = \sum_{k=1}^{c} \text{SCAT}(\mathcal{C}_k), \qquad (11.15)$$

where $\mathcal{C}_k = \{x_i : \alpha(i) = k\}$ represents the kth cluster. Similarly, the between-group scatter *of $\mathcal{X}$ can be defined as*

$$\text{SCB} = \text{SCAT}(\mathcal{X}) - \sum_{k=1}^{c} \text{SCAT}(\mathcal{C}_k), \tag{11.16}$$

so that the total scatter can be decomposed as $\text{SCT} = \text{SCW} + \text{SCB}$.

Note that if we let $x_{ki} \in \mathcal{C}_k$ represent the ith instance of cluster k, we may express the scatter of cluster k as

$$\text{SCAT}(\mathcal{C}_k) = \frac{1}{2r_k} \sum_{i=1}^{r_k} \sum_{j=1}^{r_k} d(x_{ki}, x_{kj})^2.$$

Our next result shows that the within-group scatter is an appropriate generalization of the within-group sum of squares (SSW) for non-vector metric spaces.

Proposition 11.3. *Given an inner-product space, the* SSW *given by Equation (11.11) is equivalent to the* SCW *of Equation (11.15).*

Proof. First, note that Equation (11.11) is equivalent to

$$\text{SSW} = \sum_{i=1}^{n} \left|\left| x_i - \bar{x}_{\alpha(i)} \right|\right|^2 = \sum_{k=1}^{c} \sum_{j=1}^{r_k} \left|\left| x_{kj} - \bar{x}_k \right|\right|^2$$

in the notation of Definition 11.9. Next, if we fix any particular cluster k and use the fact that $d(x, y) = ||x - y||$, we may compute the scatter of $\mathcal{C}_k$ as

$$\text{SCAT}(\mathcal{C}_k) = \frac{1}{2r_k} \sum_{i=1}^{r_k} \sum_{j=1}^{r_k} \left|\left| x_{ki} - x_{kj} \right|\right|^2$$

$$= \frac{1}{2r_k} \sum_{i=1}^{r_k} \sum_{j=1}^{r_k} \left|\left| (x_{ki} - \bar{x}_k) + (\bar{x}_k - x_{kj}) \right|\right|^2$$

$$= \sum_{i=1}^{r_k} \left|\left| x_{ki} - \bar{x}_k \right|\right|^2 + \frac{1}{r_k} \sum_{i=1}^{r_k} \sum_{j=1}^{r_k} \langle x_{ki} - \bar{x}_k, x_{kj} - \bar{x}_k \rangle.$$

We leave it as an exercise for the reader to show that the inner product on the right-hand side vanishes. By summing over the clusters, we therefore obtain the relation $\text{SCW} = \text{SSW}$, completing our result. $\square$

11.2.6 K-Medoids Clustering Algorithm for Metric Spaces

In order to generalize the K-means clustering algorithm, which operates over normed linear spaces, to a clustering algorithm that functions on an arbitrary metric space, we must first replace the within-group sum of squares optimization objective with the within-group scatter and second find an alternative for the cluster means, which require vector addition to compute.

To resolve this issue, we replace the c means $m_1, \ldots, m_c$ that are calculated in K-means with values of the actual data points, as shown in Algorithm 11.5. Notice that line 8 is now $O(n^2)$, as in order to determine the optimal m for each cluster, we must iterate through the cluster twice: for each element in a given cluster we must compute the full sum of distances, which requires summing over all the elements of the cluster. Though this algorithm is more general, it comes at the cost of computational speed.

Problems

11.1. Show that the *French-rail metric*, defined by

$$d(x, y) = \begin{cases} ||x - y||_2 & \text{if } x = \alpha y, \text{ for some } \alpha \text{ in } \mathbb{R} \\ ||x||_2 + ||y||_2 & \text{otherwise} \end{cases},$$

where $|| \cdot ||_2$ is the ℓ_2 norm (Euclidean distance to the origin) defines a metric space on $\mathbb{R}^2$. Explain the practical meaning of this metric, assuming that Paris is located at $(0,0)$, and assuming that all rail lines in France go to Paris.

Algorithm 11.5: kMedoids($\mathcal{X}, c$) – algorithm for finding c clusters over a metric space.

Input: data $\mathcal{X} = \{x_i\}_{i=1}^n$ belonging to a metric space $\mathbb{X}$;
        the number c of clusters
Output: The set $\{m_1, \ldots, m_k\}$ of cluster medoids.
1 Let $(i_1, \ldots, i_n)$ be a random shuffle of $(1, \ldots, n)$
2 Set $m_k = x_{i_k}$ for $k = 1, \ldots, c$
3 Set $\Delta = 1$
4 **while** $\Delta > 0$ **do**
5   Calculate the assignment $\alpha(i) = \arg\min_{k=1,\ldots,c} d(x_i, m_k)$
6   Let $\delta_{ik} = \mathbb{I}[\alpha(i) = k]$ and $r_k = \sum_{i=1}^{n} \delta_{ik}$
7   **for** $k = 1, \ldots, c$ **do**
8   Set $m_k' = \arg\min_{m \in C_k} \sum_{i=1}^{r_k} d(m, x_{ki})$
9   **end**
10   Set $\Delta = \sum_{k=1}^{c} d(m_k, m_k')^2$
11   Set $m_k = m_k'$, for $k = 1, \ldots, c$
12 **end**
13 **return** $(m_1, \ldots, m_k)$

11.2. Explain why the Chebyshev distance $d_\infty(x, y) = \max_j |x_j - y_j|$ in $\mathbb{R}^2$ corresponds to the distance perceived by the king on a chessboard, who can move both diagonally and horizontally and vertically, but only one step at a time.

11.3. [Bubble Sort] *Bubble sort* is an $O(n^2)$ sorting algorithm that proceeds as follows: we begin by iterating through the list that we desire to sort. We compare each item in the list with the next item, swapping them if they are out of order. Once we've reached the end of the list, the largest item will be in the final position. We then iterate through the list again, except stopping at the penultimate position, so that the largest two items are ordered at the end of the list. We continue, each time stopping one place earlier than before.

(a) Implement the bubble-sort algorithm in Python.
(b) Test the algorithm on a random array: `np.random.randint(0, 100, size=100)`.
(c) Explain why the complexity is $O(n^2)$.
(d) Modify the algorithm so that, given an integer k, the k *smallest* values in the list will be sorted at the beginning of the list. Explain why the complexity is $O(kn)$.

11.4. Show that the function d defined in Proposition 11.1 constitutes a metric.

11.5. Show that the function $|| \cdot ||$ defined in Proposition 11.2 constitutes a norm.

12

The Forest for the Trees

In this chapter, we will consider various algorithms that are rooted in *decision trees*. We first introduce decision trees, discussing how they may be grown for the cases of classification and regression, and then extend that simple model into more complex *ensemble models*, such as *random forest* and *boosted trees*. Ensemble methods constitute a straightforward way to build power machine learning algorithms from a collection of so-called *weak learners*. We will explore these concepts over the pages that follow.

12.1 Decision Trees for Classification and Regression

12.1.1 Growing Decision Trees

Decision trees, alternatively known as *rule-based methods*, are an intuitive way to learn complex patterns from data. They are based on the concept of a tree graph (Definition 8.6), with the added benefit that each node has a *rule* attached to it, that instructs us on how to partition the dataset. In this way, the output of the model is piecewise constant over a complex family of hyperrectangles that divide the feature space. Decision trees are not prescribed, but learned; the structure is uncovered from the training data, not set in advanced.

Decision Trees

Be begin by defining the notation of a *decision rule*, or, more simply, a *rule*. This definition will then allow us to formally define a decision tree.

Definition 12.1. *A* decision rule *(or* rule *or* decision criterion*) is a mapping* $r : \mathbb{D} \to \mathbb{B}$, *from the set of possible data* $\mathbb{D}$ *(relative to a given problem) to the booleans* $\mathbb{B} = \{0, 1\}$.
   Applied to a set of data $\mathcal{D}$, *we say that*

$$r(\mathcal{D}) = \{x : x \in \mathcal{D} \text{ and } r(x)\};$$

i.e.; $r(\mathcal{D})$ is the subset of data for which the decision rule is True.

Similarly, we may apply a set of decision rules $\mathcal{R}$ to a set of data $\mathcal{D}$ in the natural way:

$$\mathcal{R}(\mathcal{D}) = \{r(\mathcal{D}) : r \in \mathcal{R}\}.$$

A set of decision rules $\mathcal{R}$ is said to be mutually exclusive and exhaustive *if*

$$\mathcal{D} = \bigcup_{r \in \mathcal{R}} r(\mathcal{D}),$$

for any possible set of data $\mathcal{D} \subset \mathbb{D}$, even when counting multiplicities. Alternatively, the set $\mathcal{R}$ is mutually exclusive and exhaustive if, for any possible datum $x \in \mathbb{D}$, $r(x)$ is True for one and only one rule r in the set $\mathcal{R}$; i.e., if $\sum_{r \in \mathcal{R}} r(x) = 1$, for any x.

It is important to note that a mutually exclusive and exhaustive rule set $\mathcal{R}$ always partitions the space of feature data. Oftentimes, this partition is binary and can be expressed in the form

$$\mathcal{R} = \{r, \neg r\},$$

for some rule r, where $\neg r$ represents the negation "*not r.*" For numeric features, the rule r is typically of the form $x \geq x_0$, where x_0 is a given value. For categorical features, the rule r is typically of the form, for example, $x = \text{red}$.

Definition 12.2. *A* decision tree *is a tree graph (Definition 8.6), such that each branch node stores a set of mutually exclusive and exhaustive decision rules that map one-to-one and onto its children, and such that each leaf node stores a value.*

A classification tree *is a decision tree for which its values are categorical. A* regression tree *is a decision tree with numerical values.*

When each branch node of a decision tree has precisely two children, we call it a binary decision tree. *Otherwise, we say that the tree allows for multiway splits.*

General Algorithm for Growing Trees

Now that we have defined *what* decision trees are, we next discuss *how* to construct them.

The general algorithm for growing decision trees is given in Algorithm 12.1.

This algorithm is a high-level procedure that can be applied to both classification and regression problems. It relies on the following three functions:

- *isHomogeneous*($\mathcal{D}$): returns **True** if the data $\mathcal{D}$ are similar enough to break recursion and constitute a leaf; otherwise returns **False**;
- *getLabel*($\mathcal{D}$): returns the best predictor (label) for the data set;
- *bestSplit*($\mathcal{D}, \mathcal{F}$): returns a set $\mathcal{R}$ of mutually exclusive and exhaustive decision rules, based on the input data and feature set.

Each of the three functions, *isHomogeneous*, *getLabel*, *bestSplit*, used to grow a general decision tree have specific implementations depending on whether the problem is a classification or a regression problem.

For classification problems, *isHomogeneous* typically returns **True** if all instances in $\mathcal{D}$ have the same label. However, as this easily leads to over-fitting, it is common to instead define *isHomogeneous* to return **True** only if the set of data is homogeneous *enough*; for example, if 90% of instances belong to the same class. For regression problems, *isHomogeneous* can be defined to return **True** if the RMSE is below some critical threshold.

For classification problems, *getLabel* typically returns the *mode*; i.e., the most common label in the input data. For regression problems, *getLabel* is typically defined to return the mean or median value of the target variable.

We will discuss the details of how the *bestSplit* function works in our next paragraph. Typically this method selects a single feature to perform the split on. If the feature is numeric, the split is typically chosen to be binary: $\mathcal{R} = \{x \geq x_0, x < x_0\}$, for some optimal value x_0. If the feature is categorical, and we are allowing for multiway splits, the split is typically all possible values of a single categorical variable; for example, $\mathcal{R} = \{x = $ red$, x = $ green$, x = $ blue$\}$, assuming the only possible colors are red, green, and blue (RGB color space). Multiway splits, however, suffer from a serious

---

Algorithm 12.1: growTree($\mathcal{D}, \mathcal{F}$) – algorithm for building deci-sion trees.

Input: data $\mathcal{D}$;
            a set of features $\mathcal{F}$
Output: A decision tree T.

1 **if** isHomogeneous ($\mathcal{D}$) **then**
2 $\quad$ **return** getLabel ($\mathcal{D}$)
3 **end**
4 $\mathcal{R} = $ bestSplit ($\mathcal{D}, \mathcal{F}$)
5 $map = $ Dict ()
6 **for** r in $\mathcal{R}$ **do**
7 $\quad$ **if** $r(\mathcal{D}) = \emptyset$ **then**
8 $\quad\quad$ $map[r] = $ getLabel ($\mathcal{D}$)
9 $\quad$ **else**
10 $\quad\quad$ $map[r] = $ growTree ($r(\mathcal{D})$, $\mathcal{F}$)
11 $\quad$ **end**
12 **end**
13 **return** map

---

drawback: they tend to select features with higher cardinality first, thinning out the data set too soon. To remedy this, it is much more common to consider binary trees, which only allow for a binary partition. For binary trees, an example split would be $\mathcal{R} = \{x = \text{red}, x \neq \text{red}\}$.

Finally, we note that Algorithm 12.1 uses recursion[1] to construct a decision tree, as it makes recursive calls to itself (line 10) until it achieves the homogeneity condition (line 1).

Finding the Best Split

Next, we explore in depth the *bestSplit* function used to grow our decision trees. Both classification and regression trees rely on a certain function known as an *impurity measure* that is used to judge how good any particular split is. Typically an impurity of zero implies that the data are completely homogeneous. Thus, our objective can be restated as an optimization problem: find the split that minimizes impurity. We will restrict our attention to the usual case of a univariate split, but which we mean a mutually exclusive and exhaustive rule set that partitions the data based on any *one* feature, though it is possible to define higher dimensional splits.

The algorithm for finding the best possible split is given in Algorithm 12.2. This algorithm is specifically written to locate the optimal *binary* split, without requiring that categorical variables were previously one-hot encoded (see Section 6.2). To instead locate the optimal multiway split, the for-loop on line 5 may be removed and line 6 may be replaced with the alternative

$$\mathcal{R}' = \{x[f] = l : l \in \texttt{getLevels}(f)\}.$$

However, this is typically ill advised.

On line 13, we solving a separate optimization problem, to find the value x_0 for the numerical feature f that minimizes the post-split impurity. The standard approach is to simply iterate over the observed values of the feature f and select the value which minimizes impurity. This out create a new for-loop for numerical features. There is a simplification for classification problems, in that only such that only values x_0 for which the class label *switches* need be included, as it can be shown that the value x_0 that minimizes impurity cannot occur if the label is the same on its immediate left and right. We leave line 14 as an abstract minimization, as one-dimensional line search algorithms might be more suitable than a brute force approach, especially when training models with large sets of data. One such method, golden-section search, is given in Algorithm 12.3; though for more details and other line search methods, see Chong and Zak [2008].

---

[1] see **recursion** in the index.

Impurity

There are several possible impurity measures for classification that are commonly used; though, in practice, they are quite similar to each other.

Definition 12.3 (Impurity Measures for Classification). *For a classification problem with $c \geq 2$ classes and a data set $\mathcal{D}$, let*

$$\hat{p}_k = \frac{1}{n} \sum_{i=1}^{n} \mathbb{I}[y_i = \mathcal{C}_k], \tag{12.1}$$

represent the fraction of instances in $\mathcal{D}$ belonging to the kth class $\mathcal{C}_k$. We may then defining the following impurity measures:

1. Misclassification Error—*the observed error rate when we use the majority class as the predictor:*

$$\mathrm{Imp}(\mathcal{D}) = 1 - \max(\hat{p}_1, \ldots, \hat{p}_c); \tag{12.2}$$

Algorithm 12.2: bestSplit$(\mathcal{D}, \mathcal{F})$ – algorithm for determining best binary split.

Input: data $\mathcal{D}$;
        a set of features $\mathcal{F}$
Output: A mutually exclusive and exhaustive set of decision rules based
           on a single feature $\mathcal{R}$.

1 $I = \mathtt{getImpurity}(\mathcal{D})$
2 $\mathcal{R} = \emptyset$
3 **for** f in $\mathcal{F}$ **do**
4   **if** **type**$(f) = categorical$ **then**
5   **for** l in $\mathtt{getLevels}(f)$ **do**
6           $\mathcal{R}' = \{x[f] = l, x[f] \neq l\}$
7           $I' = \mathtt{getImpurity}(\mathcal{R}'(\mathcal{D}))$
8   **if** $I' < I$ **then**
9              $I, \mathcal{R} = I', \mathcal{R}'$
10   **end**
11   **end**
12   **else**
13        $x_0 = \mathtt{minimize}(\mathtt{getImpurity}(\mathcal{D}[x[f] >= x_0], \mathcal{D}[x[f] < x_0]))$
14        $\mathcal{R}' = \{x[f] \geq x_0, x[f] < x_0\}$
15        $I' = \mathtt{getImpurity}(\mathcal{R}'(\mathcal{D}))$
16   **if** $I' < I$ **then**
17           $I, \mathcal{R} = I', \mathcal{R}'$
18   **end**
19   **end**
20 **end**
21 **return** $\mathcal{R}$

---

2. Gini Index—*a measure of variance:*

$$\text{Imp}(\mathcal{D}) = \sum_{k=1}^{c} \hat{p}_k(1 - \hat{p}_k); \tag{12.3}$$

3. entropy—*the expected information (in bits) in learning the label of a randomly selected example:*

$$\text{Imp}(\mathcal{D}) = -\sum_{k=1}^{c} \hat{p}_k \log_2 \hat{p}_k. \tag{12.4}$$

Often, we are interested in a two-class problem, we can simplify matters by using the probability of the positive label

$$\hat{p} = \hat{p}_1 = \frac{1}{n} \sum_{i=1}^{n} \mathbb{I}[y_i = 1], \tag{12.5}$$

since the observed probability of the negative labels is given by $\hat{p}_0 = 1 - \hat{p}_1$, so that Equations (12.2)–(12.4) reduce to the expressions

$$\text{Imp}(\mathcal{D}) = \min(\hat{p}, 1 - \hat{p}), \tag{12.6}$$
$$\text{Imp}(\mathcal{D}) = 2\hat{p}(1 - \hat{p}), \tag{12.7}$$
$$\text{Imp}(\mathcal{D}) = -\hat{p} \log_2 \hat{p} - (1 - \hat{p}) \log_2(1 - \hat{p}), \tag{12.8}$$

respectively. (For a two-class problem, $\hat{p}_2 = 1 - \hat{p}_1$, so that we may drop the subscripts altogether, due to symmetry.) These impurity measures for the binary classification problem are plotted in Figure 12.1.

For regression problems, it is common to use the mean regression squared error MSR as an impurity measure:

$$\text{Imp}(\mathcal{D}) = \frac{1}{n} \sum_{i=1}^{n} (y_i - \bar{y})^2, \tag{12.9}$$

where $n = |\mathcal{D}|$, y represents the target variable, and $\bar{y}$ represents the sample mean. Alternatively, the median may replace the mean in the above formula, or a general MSE can be used (Equation (6.2)), for any other predictor for the data set.

We further require an extension to the above definitions, which will allow us to define impurity for a set of data sets.

Definition 12.4. *Any impurity measure* $\text{Imp}(\mathcal{D})$ *may be applied to a set of datasets,* $\{\mathcal{D}_1, \ldots, \mathcal{D}_m\}$ *via the linear relationship*

$$\text{Imp}(\mathcal{D}_1, \ldots, \mathcal{D}_m) = \sum_{j=1}^{m} \frac{|\mathcal{D}_j|}{|\mathcal{D}|} \text{Imp}(\mathcal{D}_j), \tag{12.10}$$

where $\mathcal{D}$ *is the set obtained by combining each subset, such that* $|\mathcal{D}| = \sum_{j=1}^{m} |\mathcal{D}_j|$.

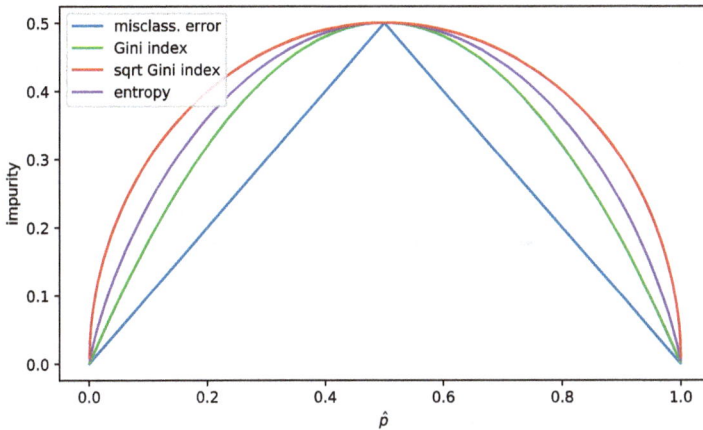

Fig. 12.1: Impurity measures for binary classification. The square-root of the Gini index and entropy are each scaled so they pass through the point $(1/2, 1/2)$.

Thus, the impurity of a collection of datasets is simply the weighted average of the individual impurities. This is the definition used in lines 7 and 13 of Algorithm 12.2, which relies on one of our individual impurity measures for either classification Equations (12.2)–(12.4) or regression Equation (12.9).

Golden-section Search

The golden-section search algorithm, due to Kiefer [1953], for locating a minimum value of a fucntion f on an interval $[a, b]$ is given in Algorithm 12.3. It is based on the golden ratio $\varphi = (1 + \sqrt{5})/2 \approx 1.61803....$, and the values $r = 2 - \varphi \approx 0.382$ and $s = \varphi - 1 \approx 0.618$, which add to unity. When we divide the unit interval into two subintervals of length r and s, the ratio $1 : s$ is equal to the ratio $s : r$, so that $r = s^2$. An improvement on this algorithm, known as Fibonacci search, can be made by varying the scaling between successive iterations, based on ratios of Fibonacci numbers. We will, however, focus on golden-section search for simplicity.

On lines 4 and 5, as well as 11 and 16, we are constructing two intermediary points $x_2, x_3 \in [x_1, x_4]$ such that

$$x_2 = x_1 + r(x_4 - x_1)$$
$$x_3 = x_4 - r(x_4 - x_1).$$

Thus, the three subintervals $[x_1, x_2]$, $[x_2, x_3]$, and $[x_3, x_4]$, relative to the overall interval, are in the ratio $r : rs : r$. It can easily be shown that $rs =$

$1 - 2r$ by using the property $\varphi^2 = \varphi + 1$. These special properties of golden ratios allow us to iteratively divide our total search area, each time getting closer to the true minimum, as shown in Figure 12.2. For additional details on the algorithm, as well as a discussion on other optimization routines, see, Chong and Zak [2008].

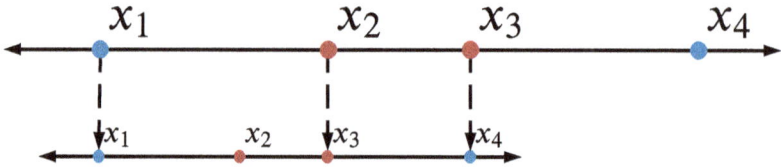

Fig. 12.2: A single iteration of the golden-section search algorithm, for $x_2 < x_3$. The initial definitions of x_1, x_2, x_3, and x_4 are on the top line, whereas the updated definitions are shown on the bottom line.

Algorithm 12.3: goldenSectionSearch – algorithm for minimizing univariate function on interval, using the golden ratio φ.

Input: a unimodal, single variable function f;
       endpoints a, b
Output: An approximate value $x \in [a, b]$ that minimizes f.

1 $\varphi = 1.61803...$
2 $r, s = 2 - \varphi, \varphi - 1$
3 $x_1, x_4 = a, b$
4 $x_2 = sx_1 + rx_4$
5 $x_3 = rx_1 + sx_4$
6 $f_1, f_2, f_3, f_4 = f(x_1), f(x_2), f(x_3), f(x_4)$
7 **while not** checkConverged$(x_1, x_2, x_3, x_4, f_1, f_2, f_3, f_4)$ **do**
8 $\quad$ **if** $\min(f_1, f_2) < \min(f_3, f_4)$ **then**
9 $\quad\quad$ $x_4, f_4 = x_3, f_3$
10 $\quad\quad$ $x_3, f_3 = x_2, f_2$
11 $\quad\quad$ $x_2 = sx_1 + rx_4$
12 $\quad\quad$ $f_2 = f(x_2)$
13 $\quad$ **else**
14 $\quad\quad$ $x_1, f_1 = x_2, f_2$
15 $\quad\quad$ $x_2, f_2 = x_3, f_3$
16 $\quad\quad$ $x_3 = rx_1 + sx_4$
17 $\quad\quad$ $f_3 = f(x_3)$
18 $\quad$ **end**
19 **end**
20 **return** the value in (x_1, x_2, x_3, x_4) corresponding to the minimum (f_1, f_2, f_3, f_4)

A few notes regarding the algorithm: the function *checkConverged* returns `True` if the values of f are sufficiently close. This can be customized as needed. Also, each successive iteration through the while loop requires only a single call to the function f. Furthermore, each successive iteration constructs an $r \approx 38.2\%$ smaller interval $[x_1, x_4]$ than the preceding one, while maintaining the $r : rs : r$ ratio of the three subdivisions. Each iteration either "zooms left" or "zooms right" depending on whether f_1 or f_2 is smaller than f_3 and f_4. If the minimum value is at x_0 or x_1, the new interval is $[x_1, x_3]$; otherwise it is $[x_2, x_4]$. Figure 12.2 depicts an example of "zooming left," corresponding to lines 9–12 of Algorithm 12.3.

When applying the algorithm to find the optimal split point for a continuous variable, we can use the minimum and maximum value of our continuous feature variable as a and b, respectively. However, if we fear our data might be susceptible to outliers, erroneous or suspicious values that live outside the fold, we can generally use, e.g., the 10th and 90th percentiles for our initial cutoff. If the optimal point is at one of our initial endpoints, we can then go back and rerun the algorithm for the outer region we initially excluded.

Finally, we note that the nth iteration of the algorithm results in a line segment that is

$$(\varphi - 1)^n \approx 0.618^n$$

the size of the original interval. We can therefore reduce the overall range by 90% with only 5 iterations of the algorithm, or by 99% with only 10 iterations. This allows us to find our optimal split point for a continuous variable by making only around a dozen calls to the impurity function, which is typically a variance calculation. This is vastly preferred over the brute-force approach, which requires performing the calculation for each possible split, especially for training sets in the millions or beyond.

A Python implementation of the golden-section search algorithm is given in Code Block 12.1. It follows directly from Algorithm 12.3.

12.1.2 Programming Decision Trees

We conclude this section with a brief discussion of an implementation of our decision tree algorithms in Python. Before we begin to implement the decision tree algorithm, we first define a `Rule` that can store and easily implement a single decision criterion. This code is shown in Code Block 12.2. A `Rule` object is initialized by a triple consisting of a feature name, a symbol (e.g., $\geq$ or $=$), and a value. We use the `operator` package in order to define a lookup table (Python dictionary) for the admissible symbols to their operators. This is a general implementation of the concept of a criterion rule that applies to both categorical features (using the symbols `'=='` and `'!='`) and numeric features (using the symbols `'>='` and `'<'`). We also defined the `__str__` method, so that we can `print(rule)` to receive an easily readable representation of the rule.

```python
def goldenSearch(f, a, b, max_iter=10, err_tol=1e-2):
    phi = (1 + 5**0.5) / 2 # golden ratio
    r, s = 2 - phi, phi - 1
    x1, x4 = a, b
    x2 = s*x1 + r*x4
    x3 = r*x1 + s*x4
    f1, f2, f3, f4 = f(x1), f(x2), f(x3), f(x4)
    converged = False
    i = 0
    while not converged:
        i += 1
        if min(f1, f2) < min(f3, f4):
            x4, f4 = x3, f3
            x3, f3 = x2, f2
            x2 = s*x1 + r*x4
            f2 = f(x2)
        else:
            x1, f1 = x2, f2
            x2, f2 = x3, f3
            x3 = r*x1 + s*x4
            f3 = f(x3)
        fmin, fmax = min(f1, f2, f3, f4), max(f1, f2, f3, f4)
        rel_err = (fmax - fmin) / fmin if fmin > 0 else 0
        converged = (i >= max_iter) or (rel_err < err_tol)
        if i >= max_iter:
            print("MAXIMUM ITERATIONS EXCEEDED")
    x = [x1, x2, x3, x4]
    f = [f1, f2, f3, f4]
    return x[np.argmin(f)]
```

Code Block 12.1: Golden-section search algorithm.

In order to apply the rule, we provide the method getMask, which returns a mask over the index of our dataframe. A *mask* is an array of booleans (True and False) that can be used to easily access slices of a given dataframe or array. We will make extensive use of the getMask functionality in our implementation of a decision tree.

We now have everything we need to construct a DecisionTree class in Python, which is shown in Code Blocks 12.3–12.6. Our DecisionTree class is subclassed from the Model abstract class, which we defined in Code Block 6.22.

The __init__ method, shown in Code Block 12.3, handles the processing for all the model parameters. First, we require that the keyword argument type is defined whenever the class is implemented, taking on a value in classification or regression, so that the algorithm knows whether the target variable is categorical or numerical, respectively.

```
1   import operator
2   class Rule:
3
4       def __init__(self, feature, symbol, value=None, cat=False):
5           self.ops = {'>': operator.gt,
6                       '<': operator.lt,
7                       '>=': operator.ge,
8                       '<=': operator.le,
9                       '==': operator.eq,
10                      '!=': operator.ne}
11          assert symbol in self.ops
12          self.feature = feature
13          self.symbol = symbol
14          self.value = value
15          self.name = self.feature + (':' + str(self.value)) * cat
16
17      def __str__(self):
18          return f"{self.feature} {self.symbol} {self.value}"
19
20      def getMask(self, df):
21          if self.value is None:
22              return np.ones(len(df)).astype(bool)
23          return self.ops[self.symbol](df[self.feature], self.value)
24
25      def getFeatureName(self, with_value=False):
26          return self.feature + (':' + str(self.value)) * with_value
```

Code Block 12.2: Implementation of a `Rule` class that stores and executes a decision criterion.

In addition, we have a few other goodies: `impurity`, which defines the impurity metric; `max_depth`, which defines a possible stopping criterion; `depth`, which is used in the `train` method to keep track of the current depth of the tree; `min_samples_split`, another stopping criterion; `min_impurity`, for classification, the minimum percentage (e.g., 0.92) to determine that a dataset is homogeneous. We also have two important keyword arguments that must be passed into the init method: `cat_features` and `num_features`. These specify the names of the categorical and numeric features to be used in model training, and should match the column names in the Pandas dataframe passed in to the `train` and `predict` methods.

We also define the `__str__` method, which prints a representation of the full set of decision rules, once the tree has been trained.

Moving on, we find that Code Block 12.4 defines the main impurity measures, entropy, Gini index, and MSE, and a `_getImpurity` method, which

```python
class DecisionTree(Model):

    def __init__(self, **kwargs):
        type_error = "Keyword argument 'type' must be set to
            classification or regression"
        assert 'type' in kwargs, type_error
        assert kwargs['type'] in ['classification', 'regression'],
            type_error
        self.type = kwargs['type']

        if self.type == 'regression':
            _allowed = ['mse']
            self.impurity = kwargs.get('impurity', 'mse')
            assert self.impurity in _allowed, f"Regression impurity
                must be in {_allowed}"
        else:
            _allowed = ['gini', 'entropy']
            self.impurity = kwargs.get('impurity', 'gini')
            assert self.impurity in _allowed, f"Classification
                impurity must be in {_allowed}"

        self.max_depth = kwargs.get('max_depth', 5)
        self.depth = kwargs.get('depth', 0) + int('depth' in kwargs)
        self.min_samples_split = kwargs.get('min_samples_split', 0)
        self.min_purity = kwargs.get('min_purity', 1)
        assert self.min_purity <= 1, f"min_purity {self.min_purity}
            must be <= 1"
        self.subspace_dim = kwargs.get('subspace_dim', 0)
        self.cat_features = kwargs.get('cat_features', [])
        self.num_features = kwargs.get('num_features', [])
        self.is_leaf = False
        self.value = None

    def __str__(self):
        preamble = '|   ' * self.depth
        str_out = ''
        if self.is_leaf:
            str_out += preamble + f"|--- value: {self.value}\n"
        else:
            for rule, tree in zip(self.rules, self.children):
                str_out += preamble + f"|--- {str(rule)}\n" +
                    str(tree)
        return str_out
    # continued ...
```

Code Block 12.3: DecisionTree implementation (part 1).

```python
# ... continued
def _gini(self, y, weights=None):
    classes = np.unique(y)
    impurity = 0
    for c in classes:
        p = np.average(y==c, weights=weights)
        impurity += p * (1 - p)
    return impurity

def _entropy(self, y, weights=None):
    classes = np.unique(y)
    impurity = 0
    for c in classes:
        p = np.average(y==c, weights=weights)
        impurity -= p * np.log2(p)
    return impurity

def _mse(self, y, weights=None):
    return np.average( (y - np.average(y, weights=weights,
        axis=0))**2, weights=weights, axis=0)

def _getImpurity(self, y, masks=None, weights=None):
    imp_funcs = {'mse': self._mse,
                 'gini': self._gini,
                 'entropy': self._entropy}
    imp_func = imp_funcs[self.impurity]
    if not masks:
        return imp_func(y, weights=weights)
    impurities, impurity_weights = [], []
    weights = np.ones(len(y)) if weights is None else weights
    for mask in masks:
        if sum(mask) == 0:
            continue
        impurities.append(imp_func(y[mask],
            weights=weights[mask]))
        impurity_weights.append(sum(weights[mask]))
    return np.average(impurities, weights=impurity_weights)
# continued ...
```

Code Block 12.4: `DecisionTree` implementation (part 2).

takes as arguments any number of datasets, allowing for both binary and multiway splits.

Code Block 12.5 defines the methods to get the best value for a given dataset and whether or not a dataset is homogeneous. It also implements the _bestSplit algorithm. The _bestSplit algorithm leverages our Rule class (Code Block 12.2) as well as the goldenSearch algorithm (Code Block 12.1) for numeric features.

Note that all of the methods defined so far are preceded with a single underscore, signifying the intent that they are used internally, and not outside of the class definition.

Finally, our main train and predict methods are implemented in Code Block 12.6. Note the use of recursion in the train method: each decision tree begets two new decision trees (or more, if we were to redefine the _bestSplit to allow for multiway splits), stored as the children of the tree. The process stops when the homogeneity condition is reached, or when we reach a max_depth or the data thins below the size of min_samples_split.

Once a decision tree has been trained, we can call the predict method to get a vector of predictions, which are constructed using vector operations (via our masks) and the recursive structure of the full tree.

Finally, let's apply our decision tree algorithm to the class *iris data set*, available from sklearn. The code is shown in Code Block 12.7. The output of the print(T) command is shown in Code Block 12.8. The tree of maximum depth 3 is shown in Figure 12.3. With a maximum tree depth of 5, we achieve 100% accuracy on the training set. Though, with a tree depth of 2 or 3, we still achieve 96% or 97% accuracy.

12.1.3 Decision Trees for Probability Estimation

Finally, we note a connection between the Gini index

12.1.4 Tree Pruning

Decision trees are prone to overfit their training data, especially if they are allowed to grow to an arbitrary depth. This is because, given an endless supply of splits, a decision tree can always "memorize" the training set, but eventually taking into account all possible (realized) combinations of the input set. To remedy this situation, we can *prune* our initial decision tree, by removing branches that do not generalize well. Before proceeding, note that this method is typically only applied when one trains a single tree, and is not used for the ensemble methods we will discuss for the remainder of the chapter. This section may therefore be omitted by the anxious reader without loss in continuity.

The tree pruning algorithm is shown in Algorithm 12.4. Here, we define $T_N = \text{subTree}(T, N)$ as the subtree of tree T rooted at node N. Then we define $T_N(\mathcal{D}_N)$ as the set of labels for $\mathcal{D}_N$ based on the leaves of subtree T_N.

```python
1   # ... continued
2   def _getValue(self, y, weights=None):
3       if self.type == 'classification':
4           prob = {k: v / len(y) for k, v in Counter(y).items()}
5           return max(prob, key=prob.get), prob
6       else:
7           return np.average(y, weights=weights, axis=0), None
8
9   def _isHomogeneous(self, y, weights=None):
10      if self.type == 'classification':
11          mode, _ = self._getValue(y, weights=weights)
12          return np.average(y==mode, weights=weights) >= \
                  self.min_purity
13      else:
14          return False
15
16  def _bestSplit(self, df, y, cat_features_dict, num_features,
        weights=None):
17      weights = weights if weights else np.ones(len(df))
18      min_impurity = imp_0 = self._getImpurity(y)
19      rules = []
20      for feature, levels in cat_features_dict.items():
21          for level in levels:
22              _rules = [Rule(feature, sym, level, cat=True) for sym
                      in ['==', '!=']]
23              masks = [rule.getMask(df) for rule in _rules]
24              impurity = self._getImpurity(y, masks, weights)
25              if impurity < min_impurity:
26                  min_impurity = impurity
27                  rules = _rules
28      for feature in num_features:
29          a, b = df[feature].min(), df[feature].max()
30          x = goldenSearch(lambda x: self._getImpurity(y,
                  masks=[(df[feature] < x).tolist(), (df[feature] >=
                  x).tolist()], weights=weights), a, b)
31          _rules = [Rule(feature, '>=', x), Rule(feature, '<', x)]
32          masks = [rule.getMask(df) for rule in _rules]
33          impurity = self._getImpurity(y, masks, weights)
34          if impurity < min_impurity:
35              min_impurity = impurity
36              rules = _rules
37      return rules, imp_0 - min_impurity
38
39  def _getFeatureSpace(self):
40      return self.cat_features_dict, self.num_features
41  # continued ...
```

Code Block 12.5: DecisionTree implementation (part 3).

```python
# ... continued
def train(self, df, y, weights=None):
    self.value, self.prob = self._getValue(y)
    if not hasattr(self, 'cat_features_dict'):
        self.cat_features_dict = {feature:
            np.unique(df[feature]).tolist() for feature in
            self.cat_features}
        self.n_features = len(self.num_features) + sum([len(x)
            for _, x in self.cat_features_dict.items()])
    if not hasattr(self, 'vector_dim'):
        self.vector_dim = y.shape[1] if y.ndim == 2 else None
    if self._isHomogeneous(y) or (self.depth >= self.max_depth)
        or (len(y) < self.min_samples_split):
        self.is_leaf = True
        return
    weights = np.ones(len(df)) if weights is None else weights
    cat_features_dict, num_features = self._getFeatureSpace()
    self.rules, self.impurity_reduction = self._bestSplit(df, y,
        cat_features_dict, num_features, weights=weights)
    self.children = [DecisionTree(**self.__dict__) for rule in
        self.rules]
    for rule, child in zip(self.rules, self.children):
        mask = rule.getMask(df)
        child.train(df[mask], y[mask], weights=weights[mask])

def predict(self, df, get_prob=False):
    if self.is_leaf:
        return self.prob if get_prob else self.value
    if self.vector_dim:
        y_hat = np.zeros((len(df), self.vector_dim))
    elif isinstance(self.value, (int, float)):
        y_hat = np.zeros(len(df))
    else:
        y_hat = np.array(['' for i in range(len(df))],
            dtype='object')
    for rule, child in zip(self.rules, self.children):
        mask = rule.getMask(df)
        y_hat[mask] = child.predict(df[mask], get_prob=get_prob)
    return y_hat
```

Code Block 12.6: DecisionTree implementation (part 4).

```
1  from sklearn.datasets import load_iris
2  data = load_iris()
3  print(data.DESCR)
4  X = data.data
5  y = data.target
6  df = DataFrame(X, columns = data.feature_names)
7  y = np.array(list(map(lambda x: data.target_names[x], y)))
8
9  T = DecisionTree(type='classification', num_features=['sepal length
       (cm)', 'sepal width (cm)', 'petal length (cm)', 'petal width
       (cm)'], max_depth=5, min_samples_split=0)
10 T.train(df, y)
11 print(T)
12 np.sum(T.predict(df) == y) / len(y) # 1.00
```

Code Block 12.7: Training our decision tree on the iris data set.

```
1  |--- petal length (cm) >= 2.9544512279404422
2  |   |--- petal width (cm) >= 1.791796067500631
3  |   |   |--- petal length (cm) >= 4.989356881873896
4  |   |   |   |--- value: virginica
5  |   |   |--- petal length (cm) < 4.989356881873896
6  |   |   |   |--- sepal width (cm) >= 3.0090169943749476
7  |   |   |   |   |--- value: versicolor
8  |   |   |   |--- sepal width (cm) < 3.0090169943749476
9  |   |   |   |   |--- value: virginica
10 |   |--- petal width (cm) < 1.791796067500631
11 |   |   |--- petal length (cm) >= 5.584359886491461
12 |   |   |   |--- value: virginica
13 |   |   |--- petal length (cm) < 5.584359886491461
14 |   |   |   |--- petal length (cm) >= 4.9829710109982335
15 |   |   |   |   |--- petal width (cm) >= 1.547213595499958
16 |   |   |   |   |   |--- value: versicolor
17 |   |   |   |   |--- petal width (cm) < 1.547213595499958
18 |   |   |   |   |   |--- value: virginica
19 |   |   |   |--- petal length (cm) < 4.9829710109982335
20 |   |   |   |   |--- petal width (cm) >= 1.7
21 |   |   |   |   |   |--- value: virginica
22 |   |   |   |   |--- petal width (cm) < 1.7
23 |   |   |   |   |   |--- value: versicolor
24 |--- petal length (cm) < 2.9544512279404422
25 |   |--- value: setosa
```

Code Block 12.8: Trained decision diagram for classifying the iris data set.

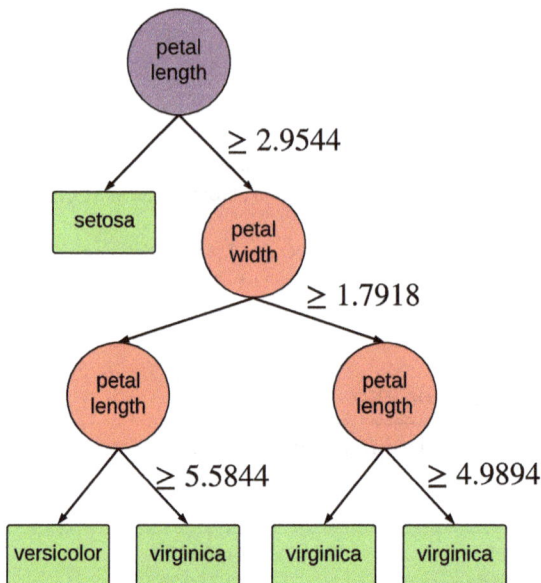

Fig. 12.3: A tree of depth 3 trained on the iris data set.

In this way, the act of pruning is tantamount to removing all subtrees that are detrimental to the overall performance.

12.2 Random Forests

In this section, we discuss a powerful extension to the decision tree algorithm. The main idea is that we will train many individual decision trees

---

Algorithm 12.4: pruneTree($T, \mathcal{D}$) – algorithm to prune a decision tree.

Input: tree T;
　　　　data $\mathcal{D}$
Output: A pruned decision tree T'.

1 **for** *each internal node N of T, starting at bottom* **do**
2 　　$T_N = $ subTree(T, N)
3 　　$\mathcal{D}_N = \{x \in \mathcal{D} : x$ is covered by $N\}$
4 　　**if** getImpurity$(T_N(\mathcal{D}_N)) > $ getImpurity$(\mathcal{D}_N)$ **then**
5 　　　　| replace T_N in T with a leaf labelled by getLabel$(\mathcal{D}_N)$
6 　　**end**
7 **end**
8 **return** T

---

and pool their collective wisdom together under a single model. Such algorithms are known as *ensemble methods*, and they are very powerful.

12.2.1 Ensemble Methods

We begin with a brief introduction to *ensembles*, as defined below.

Definition 12.5. *An* ensemble model *(or* voting assembly*) is a finite collection of m pairs,*

$$\mathcal{E} = \{(M_j, w_j)\}_{j=1}^{m}, \tag{12.11}$$

such that each M_j is a model over the same feature and target space; such that the values w_j are weights, possessing the property that $w_j > 0$; and such that the ensemble model itself constitutes a function from the feature space to the target space that combines the individual predictions, for an arbitrary input, with their respective weights to return a unique output.

An ensemble model is normalized *if the weights are constrained by the normalization condition $\sum_{j=1}^{m} w_j = 1$; otherwise it is said to be* unnormalized.*

For regression problems, we typically define the functionality of an ensemble model as the weighted average

$$\mathcal{E}(x) = \sum_{i=j}^{m} w_j M_j(x). \tag{12.12}$$

For classification problems, on the other hand, we can define the functionality of an ensemble method via the *voting scheme*

$$\mathcal{E}(x) = \arg\max_{c \in \mathcal{C}} \left\{ c \rightarrow \sum_{j=1}^{m} w_j \mathbb{I}[M_j(x) = c] \right\}, \tag{12.13}$$

where $\mathcal{C}$ is the set of classes for the target variable. In other words, we group the individual models based on their predicted outputs, and allow each model to vote with weight w_j. The output class with the most votes wins. In Equation (12.13), we use the brackets to represent an array, indexed by the set of target classes $\mathcal{C}$; the argmax function therefore yields the *key c* corresponding to the largest *value* $\sum_{j=1}^{m} w_j \mathbb{I}[M_j(x) = c]$.

Finally, we note that ensembles are *additive* in the following sense.

Definition 12.6. *We define the* direct sum *of two ensembles $\mathcal{E} = \{(M_i, w_i)\}_{i=1}^{n}$ and $\mathcal{E}' = \{(N_j, v_j)\}_{j=1}^{m}$ as the new ensemble*

$$\mathcal{E} \oplus \mathcal{E}' = \{(M_i, w_i)\}_{i=1}^{n} \cup \{(N_j, v_j)\}_{j=1}^{m},$$

renormalized so that $\sum_{i=1}^{n} w_i + \sum_{j=1}^{m} v_j = 1$, where appropriate.

A basic ensemble abstract base class is constructed in Code Block 12.9, from which we can derive many ensemble children, by overriding the behavior of the _getModel and _trainNext methods. We will see several examples of this later in the chapter.

For the remainder of this chapter, we will explore various methods for constructing ensemble models, whose individual constituents are decision trees. Such methods are usually referred to as *forest methods*, as they represent a collection of trees.

12.2.2 Bagged Forests

The simplest way to generate a forest from some trees is to *bag* them. Here, *bagging* is a term that is short for *bootstrap aggregating*. The bagging algorithm is given in Algorithm 12.5, where it is expressed as an ensemble over a general learning algorithm $\mathscr{A}$, not limited to the decision tree algorithm. This basic algorithm is due to Breiman [1966]. Formally, we can define a bagged forest as follows.

Definition 12.7. *A* bagged forest *is a model trained using the bootstrap-aggregation Algorithm 12.5 that uses a decision tree (Algorithms 12.1 and 12.2) as the input algorithm $\mathscr{A}$.*

In Algorithm 12.5, we assume the learning algorithm $\mathscr{A}$ is a function from the data space $\mathbb{D}$, with a set of features from $\mathbb{F}$, to the space of predictive models $\mathbb{M}$; i.e., it is a function that takes a set of data as input and returns a predictive model as an output. We also assume we have a simple bootstrap method getBootstrap: $\mathbb{D} \times \mathbb{Z}_+ \to \mathbb{D}$, that returns a bootstrap sample (with replacement) from an original dataset ($\mathbb{D}$) of a given size ($\mathbb{Z}_+$). In the bagging algorithm, each bootstrap sample is of the same size as the original data set. On average, each bootstrap sample will cover $(1-1/e) \approx 63.21\%$ of the original data sample, as the size of the set $n \to \infty$.

Algorithm 12.5: bagging($\mathcal{D}, \mathcal{F}, \mathscr{A}, m$) – bootstrap aggregation algorithm. For *bagged forests*, use $\mathscr{A} =$ DecisionTree. For *random forests*, use $\mathscr{A} =$ RandomDecisionTree (see Code Block 12.12).

Input: data $\mathcal{D}$;
feature set $\mathcal{F}$;
learning algorithm $\mathscr{A} : \mathbb{D} \times \mathbb{F} \to \mathbb{M}$;
the size m of the ensemble
Output: An ensemble model $\mathcal{E}$.
1 **for** j from 1 to m **do**
2 $\quad \mathcal{B} =$ getBootstrap($\mathcal{D}, |\mathcal{D}|$)
3 $\quad M_j = \mathscr{A}(\mathcal{B}, \mathcal{F})$
4 **end**
5 **return** $\mathcal{E} = \{M_j, 1\}_{j=1}^m$

```python
from tqdm import tqdm
class Ensemble(Model):

    def __init__(self, **kwargs):
        assert kwargs['type'] in ['classification', 'regression']
        self.type, self.size = kwargs['type'], kwargs.get('size', 10)
        self.normalize = kwargs.get('normalize', True)
        self.params = kwargs.copy()
        kwargs.update(kwargs.get('model_params', {}))
        self.model_params = kwargs
        self.models, self.weights = [], []
        self.Model = self._getModel()

    @abstractmethod
    def _getModel(self):
        pass

    def _trainNext(self, df, y):
        model = self.Model(**self.model_params)
        model.train(df, y)
        return model, 1

    def train(self, df, y):
        if self.type == 'classification':
            self.classes = np.sort(np.unique(y)).tolist()
        if not hasattr(self, 'vector_dim'):
            self.vector_dim = y.shape[1] if y.ndim == 2 else None
        for j in tqdm(range(self.size)):
            model, weight = self._trainNext(df, y)
            self.models.append(model)
            self.weights.append(weight)

    def predict(self, df):
        weights = [w / sum(self.weights) for w in self.weights] if \
            self.normalize else self.weights
        if self.type == 'regression':
            y_hat = np.zeros((len(df), self.vector_dim)) if \
                self.vector_dim else np.zeros(len(df))
            for model, weight in zip(self.models, weights):
                y_hat += weight * model.predict(df)
        else:
            votes = DataFrame(0, index=np.arange(len(df)),
                columns=self.classes) # n x k
            for model, weight in zip(self.models, weights):
                y_hat = model.predict(df)
                for i, j in enumerate(y_hat):
                    votes.loc[i, j] += weight
            y_hat = votes.idxmax(axis=1).values.tolist()
        return y_hat
```

Code Block 12.9: Ensemble abstract class implementation.

12.2.3 More Magic: Out-of-bag Error and Feature Importance

Two important pieces of information can be computed and stored while training bagged forests: the so-called out-of-bag error and a stack ranking of feature importances. The former serves as a viable substitute for cross validation, while the latter provides insight into which features are making the most impact on reducing model variance, and further provides a measure of their relative importance.

Out-of-bag (OOB) Error

One bonus advantage of bagging algorithms is that validation can be done for free while building the models that comprise the ensemble. This is achieved by using the so-called *out-of-bag samples*, or OOB *samples*, which we define as follows.

Definition 12.8. *Given a bootstrap sample* $\mathcal{B} \subset \mathcal{D}$ *of a data set* $\mathcal{D}$, *the* out-of-bag (OOB) samples *are the training instances that are not used in the bootstrap; i.e., the* OOB *samples are comprised of the instances in the set* $\mathcal{D} \setminus \mathcal{B}$.

Given this definition, we can next define the OOB *score*, as follows.

Definition 12.9. *Given a bagged ensemble model (Algorithm 12.5), the* out-of-bag (OOB) score *is the score obtained by applying any scoring metric (e.g.,* R^2 *or* MSE *for regression or accuracy for classification) to the predictions obtained by aggregating the out-of-bag predictions for each individual model in the ensemble.*

Note that each training instance will, in general, have a different number of OOB predictions, based on whether or not it happened to be used for each individual bootstrap. The OOB predictions are typically computed within the for-loop of Algorithm 12.5, and their aggregation is computed following the for-loop. We will discuss the Python implementation of the OOB score in during our discussion of Code Block 12.14.

Finally, we note that the OOB score for a bootstrap ensemble is equivalent, for large data sets, to the score obtained by performing a k-fold cross validation of the set. This follows as both methods are based on the same principle: apply a trained model to a randomly selected hold-out group.

Feature Importance

We are often interested in determining which features were the most vital in training a model. After all, decision trees *learn* which features are the most useful by testing out various possibilities for each split, they are not instructed by the programmer a priori which way to go. It is therefore often of interest to take a peek at which features helped the model the most, whether the model is an individual decision tree or an entire forest.

Recall that, when training a decision tree, the best split at node N is the binary decision rule $\mathcal{R}_N$ that maximizes the impurity reduction (or impurity loss)

$$\Delta \mathrm{Imp}_N = \mathrm{Imp}(\mathcal{D}_N) - \mathrm{Imp}(\mathcal{R}_N(\mathcal{D}_N))$$

across all possible features $\mathcal{F}$, where $\mathcal{D}_N$ is the data that survived to node N. We can agree that a feature that results in a greater impurity reduction should be considered more important than a feature that results in a lesser impurity reduction. Typically, important features occur closer to the root of the tree, or are features that are selected multiple times throughout the training process for the tree.

To determine the feature importance for a given tree, we can simply track the overall impurity reduction for each feature by using a map. We can then normalize the values of the given map, so that the largest value is set at unity (1.00). We can then output a ranked list of features, sorted by importance (descending), with the relative importance listed with each feature.

For bagged forests, we can simply aggregate the map of impurity reductions across all trees in the forest, prior to normalizing and returning the stack rank.

In order to calculate feature importances in Python, we can build on our `DecisionTree` class (Code Blocks 12.3–12.6) by adding a new method, `getFeatureImportance`. We will leverage the `name` attribute in the `Rule` class (Code Block 12.2), as it captures both the feature name and the level for categorical features. We will also leverage the `impurity_reduction` attribute defined in the `train` method of Code Block 12.6, which is returned by the `_bestSplit` method in Code Block 12.5. Previously, these attributes were unused.

The code for calculating feature importances is given in Code Block 12.10. We begin with the internal method `_getImpurityReduction`, which leverages the `collections` module's `Counter` class[2] to tabulate a map from feature names to impurity reductions as we iterate through each of the tree's internal nodes.

Finally, the method `getFeatureImportances`, which takes an optional argument `top`, returns the top `top` features in the form of a `panda`'s `DataFrame` with columns `feature` and `importance`, the latter scaled so that the top feature has importance of `1.00`. Optionally, the method returns the raw `Counter` object.

---

[2] The `Counter` class operates similarly to a dictionary, except it implements the `__add__` operator, in a way that adds corresponding values together when adding two `Counter` objects.

```
1  from collections import Counter
2  class DecisionTree:
3      # ... continued
4      def _getImpurityReductions(self):
5          if self.is_leaf:
6              return Counter()
7          reductions = Counter({'n_nodes': 1, self.rules[0].name:
               self.impurity_reduction})
8          for model in self.children:
9              reductions += model._getImpurityReductions()
10         return reductions
11
12     def getFeatureImportance(self, top=0, as_counter=False):
13         if not hasattr(self, 'feature_counter'):
14             self.feature_counter = self._getImpurityReductions()
15             self.n_nodes = self.feature_counter.pop('n_nodes')
16             self.feature_importances =
                   DataFrame(self.feature_counter.most_common(),
                   columns=['feature', 'importance'])
17             self.feature_importances.importance /=
                   self.feature_importances.importance.max()
18         if as_counter:
19             return self.feature_counter
20         top = len(self.feature_importances) if top == 0 else top
21         return self.feature_importances[:top]
```

Code Block 12.10: Two additional methods for the `DecisionTree` class as continued from Code Blocks 12.3–12.6.

Oftentimes, however, we are more interested in the feature importances from an ensemble method, which aggregates the impurity reductions across all models comprising the ensemble. To achieve this, we can add a new method, `getFeatureImportance` to the `Ensemble` abstract base class. This method first ensures that the base model of the ensemble (stored as `self.Model`, which, recall, is a reference to the *class* defining the base model, not an individual object) has its own `getFeatureImportance` method defined. Next, we simply loop over the models from the ensemble, adding their feature-reduction counters in the natural way. We will see an example output in Code Block 12.16.

12.2.4 Random Forests

Bagged forests suffer from one serious drawback: at the end of the day, the trees that comprise the forest turn out to be highly correlated. To remedy the situation, we often employ one simple, yet crucial, modification

```
1   class Ensemble:
2       # ... continued
3       def getFeatureImportance(self, top=0):
4           assert hasattr(self.Model, 'getFeatureImportance'), "Base
                model class must have getFeatureImportance method."
5           if not hasattr(self, 'feature_importances'):
6               self.feature_counter = Counter()
7               for model in self.models:
8                   self.feature_counter +=
                        model.getFeatureImportance(as_counter=True)
9               self.feature_importances =
                    DataFrame(self.feature_counter.most_common(),
                    columns=['feature', 'importance'])
10              self.feature_importances.importance /=
                    self.feature_importances.importance.max()
11          top = len(self.feature_importances) if top == 0 else top
12          return self.feature_importances[:top]
```

Code Block 12.11: Additional method for the **Ensemble** class as continued from Code Block 12.9.

to our bagging algorithm, that of *feature subspace sampling*, due to Ho [1995] and Ho [1998]. The idea is that, in addition to bootstrapping the data for each tree, we also also consider only a subset of features for each split when training each tree. This prevents any single feature from dominating throughout many trees in the forest. When we couple the idea of bagging together with subspace sampling, we call the resulting model a *random forest*. It is identical to the bagging algorithm, except for the fact that we also sample the feature space, for each split of each tree.

In order to formalize this concept, we first define a *random decision tree*, as follows.

Definition 12.10. *A random decision tree is trained like an ordinary decision tree (Algorithm 12.1), except that a random subset of features, $\mathcal{F}'$, of size $f < |\mathcal{F}|$, is passed into the bestSplit method on line 4. The parameter f is referred to as the (feature) subspace dimension.*

Given this definition of a random decision tree, we may now define a random forest as follows.

Definition 12.11. *A random forest is a model trained using the bootstrap-aggregation Algorithm 12.5 that uses a random decision tree (Definition 12.10) as the input algorithm $\mathscr{A}$.*

In general, a good rule of thumb (see Breiman [2001]) is to set the feature subspace dimension f as

1. $f = |\mathcal{F}|/3$, for regression problems; and
2. $f = \sqrt{|\mathcal{F}|}$, for classification problems.

However, these rules of thumb should be regarded as starting points, and the subspace dimension as a tunable model parameter.

Python Implementation: Random Decision Trees

Our definition of _getFeatureSpace in Code Block 12.5 may have at first seemed superfluous. Yet, it serves an important function now that we wish to modify our basic decision tree code (Code Blocks 12.3–12.6) to support random feature subspace selection. Leveraging the code already built for regular decision trees, we can construct a *random decision tree* class by defining a subclass and overwriting a single method. The result is shown in Code Block 12.12.

```
class RandomDecisionTree(DecisionTree):

    def _getFeatureSpace(self):
        mask = np.random.permutation([True for i in
            range(self.subspace_dim)] + [False for i in
            range(self.n_features - self.subspace_dim)]).tolist()
        cat_features_dict, num_features = {}, []
        for feature, levels in self.cat_features_dict.items():
            n_levels = len(levels)
            x = [mask.pop() for i in range(n_levels)]
            if sum(x) == 0:
                continue
            cat_features_dict[feature] = [level for test, level in
                zip(x, levels) if test]
        for feature in self.num_features:
            if mask.pop():
                num_features.append(feature)
        return cat_features_dict, num_features
```

Code Block 12.12: RandomDecisionTree implementation.

Note that, since we built our decision tree algorithm without one-hot encoding, we need to take care in how we randomly select our features. Each level of each categorical feature is a feature in its own right, as a categorical feature variable with l levels would, ordinarily, be one-hot encoded into l independent features. We get around this problem by defining a dictionary of categorical features, cat_features_dict, that has feature name as key and list of levels as value.

We can train a random decision tree on the iris data set, using `max_depth=3` and `subspace_dim=2`. The output is shown in Code Block 12.13. We see that, for this run, the tree is making different split decisions than before, due to the randomization of the feature subspaces. Moreover, we see that the right branch (sepal length $>= 5.47$) then splits on a different feature (petal width) than the left branch, which splits on petal length.

```
1   |--- sepal length (cm) >= 5.47445651649735
2   |   |--- petal width (cm) >= 1.7914681984719456
3   |   |   |--- value: virginica
4   |   |--- petal width (cm) < 1.7914681984719456
5   |   |   |--- value: versicolor
6   |--- sepal length (cm) < 5.47445651649735
7   |   |--- petal length (cm) >= 2.9856581412139334
8   |   |   |--- value: versicolor
9   |   |--- petal length (cm) < 2.9856581412139334
10  |   |   |--- value: setosa
```

Code Block 12.13: Trained `RandomDecisionTree` for the iris data set; notice the splits are different than in Code Block 12.8.

Python Implementation: Bagged and Random Forests

We can subclass the `Ensemble` abstract base class from Code Block 12.9 to implement the bagged forest algorithm, as shown in Code Block 12.14. Most of this code is dedicated to calculating the out-of-bag score. To implement a bagged forest without the out-of-bag score, we only need to override the methods `_getModel` (in order to set the base model to our `DecisionTree` class (Code Block 12.3)) and `_trainNext` (in order to get a bootstrap sample for each model), and we only need lines 12–16 of the latter method.

Note that we invoke the built-in **super** function twice, on lines 6 and 36. In this way, when we override the `__init__` and **train** methods, we do not have to reproduce *all* of the code; rather, we are construction a *wrapper* around those methods, which performs certain additional tasks before and after.

In `__init__`, we make sure that a validation metric is defined and stored as `self.oob_metric`. We use `accuracy_score` and `r2_score`, from the `sklearn.metrics` package, as default metrics for classification and regression, respectively.

In the **train** method, we initialize a structure (`self.oob_votes`) for storing the out-of-bag predictions, which we later compute within the `_trainNext` method.

```python
class BaggedForest(Ensemble):
    def __init__(self, **kwargs):
        default_metric = r2_score if kwargs['type'] == 'regression'
            else accuracy_score
        self.oob_metric = kwargs.get('oob_metric', default_metric)
        self.get_oob = kwargs.get('get_oob', False)
        self.n_jobs = kwargs.get('n_jobs', 1)
        super().__init__(**kwargs)

    def _getModel(self):
        return DecisionTree

    def _trainNext(self, df, y):
        idx = np.random.randint(0, len(df), size=len(df))
        df_boot = df.loc[idx, :].reset_index(drop=True)
        y_boot = y[idx]
        model = self.Model(**self.model_params)
        model.train(df_boot, y_boot)
        model.oob_idx = [i for i in range(len(df)) if i not in idx]
        return model, 1

    def train(self, df, y):
        super().train(df, y)
        if not self.get_oob:
            return
        if self.type == 'regression':
            oob_votes = np.zeros(y.shape)
            oob_n_votes = np.zeros(len(df))
            for model in self.models:
                y_hat = model.predict(df.loc[model.oob_idx, :])
                oob_votes[model.oob_idx] += y_hat
                oob_n_votes[model.oob_idx] += 1
            n_votes = np.fmax(1, oob_n_votes)
            if self.vector_dim:
                n_votes = n_votes.reshape((len(y), 1))
            oob_votes /= n_votes
            has_votes = oob_n_votes > 0
            oob_votes[~has_votes] = np.average(oob_votes[has_votes],
                weights=oob_n_votes[has_votes], axis=0)
        else:
            oob_votes = DataFrame(0, index=np.arange(len(df)),
                columns=self.classes)
            for model in self.models:
                y_hat = model.predict(df.loc[model.oob_idx, :])
                for i, j in zip(model.oob_idx, y_hat):
                    oob_votes.loc[i, j] += 1
            oob_votes = oob_votes.idxmax(axis=1).values.tolist()
        self.oob_score = self.oob_metric(y, oob_votes)
```

Code Block 12.14: BaggedForest implementation; with *out-of-bag score*.

In the _trainNext method, we first perform a bootstrap sample, using numpy.random.randint, train a base model using the bootstrap sample, and then compute the predictions for the out-of-bag samples, indexed by oob_idx.

Finally, in the train method, after invoking the parent class's (super's) train method, we aggregate our out-of-bag predictions and compute our out-of-bag score.

We can further subclass the BaggedForest into the RandomForest class of Code Block 12.15, by updating the model with our RandomDecisionTree class (Code Block 12.12).

```
class RandomForest(BaggedForest):

    def _getModel(self):
        return RandomDecisionTree
```

Code Block 12.15: RandomForest implementation.

We can train a simple random forest model using 100 trees, as shown in Code Block 12.16. We see that the oob_score is 93.33%, meaning that our model's prediction accuracy is expected to be around 93%. In addition, we can invoke the getFeatureImportance method to retrieve a dictionary of the feature importances. We see that petal length and petal width are the top two features, with the former being nearly twice as important as the latter. Sepal length and sepal width are a distance third and fourth, with sepal width making very little contribution to the overall model.

Finally, we illustrate a smaller random forest with twelve trees in Figure 12.4. The oob_score with only twelve trees is 91.33%. We can actually calculate the predicted labels based on this diagram alone. For a given instance, we first calculate the prediction of each of the twelve trees individually. We then tabulate the results and choose the class with the most votes.

Figure 12.5 shows the class hierarchy for the various Python classes we've constructed so far. Note that the two base models (DecisionTree and RandomDecisionTree) are inputs into our two ensemble models (BaggedForest and RandomForest).

A regression example is shown in Code Block 12.17. This dataset uses ten features to predict the target variable, which represents a quantitative measure of disease progression one year following prognosis. The relative feature importances are shown in Figure 12.6.

```
1  E = RandomForest(type='classification', num_features=['sepal length
       (cm)', 'sepal width (cm)', 'petal length (cm)', 'petal width
       (cm)'], max_depth=2, min_samples_split=10, subspace_dim=2,
       size=100, get_oob=True)
2  E.train(df, y)
3  E.oob_score # 0.9333333
4  E.getFeatureImportance()
5  # OUTPUT:
6  #    feature                importance
7  # 0  petal length (cm)      1.000000
8  # 1  petal width (cm)       0.569108
9  # 2  sepal length (cm)      0.067717
10 # 3  sepal width (cm)       0.004357
```

Code Block 12.16: OOB score and feature importances for a random-forest model with 100 trees trained from the iris data set. Variables loaded from Code Block 12.7.

```
1  data = load_diabetes()
2  print(data.DESCR)
3  X, y = data.data, data.target
4  df = DataFrame(X, columns = data.feature_names)
5  df.sex = df.sex.apply(lambda x: 'M' if x > 0 else 'F')
6  num_features = ['age', 'bmi', 'bp', 's1', 's2', 's3', 's4', 's5',
       's6']
7  E = RandomForest(type='regression', cat_features=['sex'],
8                   num_features=num_features,
9                   max_depth=3, min_samples_split=20,
10                  subspace_dim=3, size=100, get_oob=True)
11 E.train(df, y)
12 E.oob_score # 0.44129
13 dfi = E.getFeatureImportance()
14 dfi.plot('feature', kind='barh', legend=False)
15 plt.gca().invert_yaxis()
```

Code Block 12.17: Random forest and feature importance for the diabetes dataset.

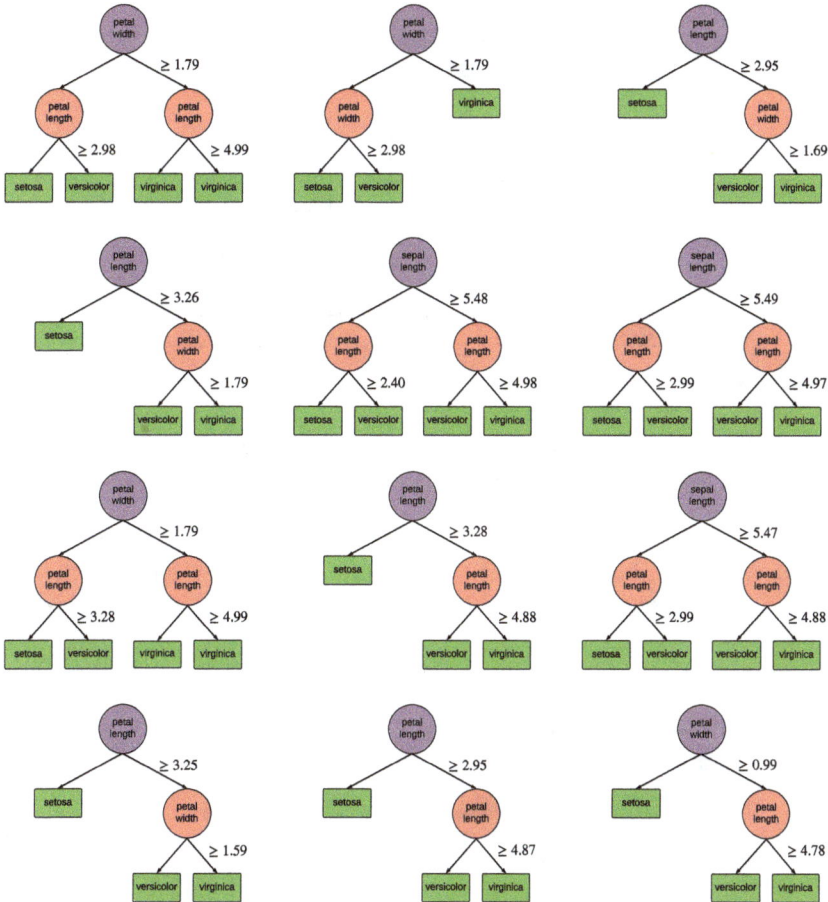

Fig. 12.4: A random forest with twelve trees, trained over the iris data set, with max depth 2.

12.3 Boosted Forests

In the previous section, we saw how boosting can be used to reduce the variance of a model, by producing an ensemble of independently trained models. In bagged forests, in particular, the trees, though independent, tend to be correlated, with top features dominating many individual trees. To remedy the correlation problem, we further introduced random forests, which use feature subspace sampling to ensure a broader variety of predictors in the forest. However, trees are still trained independently of one another.

The main idea of *boosting* is to train an ensemble of base models *in series*, so that each model can learn from the mistakes of its predecessor.

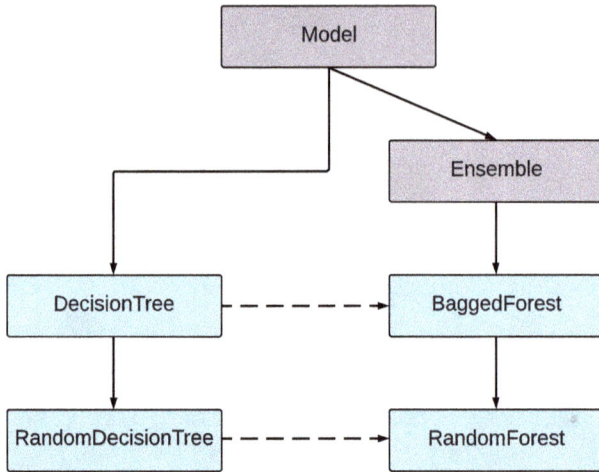

Fig. 12.5: Class hierarchy for tree and ensemble methods. Abstract classes are purple. Solid lines represent inheritance; dashed lines represent the definition of an ensemble's base class.

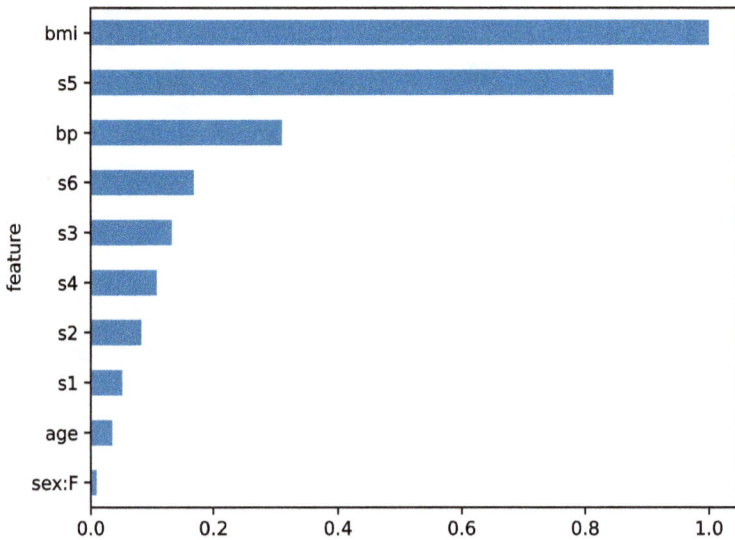

Fig. 12.6: Feature importances for the diabetes random forest example.

We will discuss two main boosting algorithms in this section: adaboost and gradient boosting.

12.3.1 AdaBoost

One of the main underlying ideas behind boosting is that many *weak learners* can be combined to form a *strong learner*. In this context, a *weak learner* is any classification model that performs slightly better than guessing class labels at random, and a *strong learner* is a model that can be tuned to produce an arbitrarily small error rate. The original boosting algorithm is due to Schapire [1990], who divided a large dataset into three subsets, training three successive weak models, and then showing that the overall error could be reduced arbitrarily by applying the method recursively. This method was improved upon by Freund and Schapire [1996], who introduced a variation known as *adaptive boosting*, or simply *AdaBoost*. The AdaBoost algorithm uses the entire dataset with each iteration, and can train an arbitrary number of base learners. We will discuss the original form of AdaBoost, known as AdaBoost.M1, which is shown in Algorithm 12.6.

Recall that bagged forests and random forests each train an independent collection of trees (or, more generally, base learners), with the random forest improvement helping to decorrelate the individual trees, providing a greater diversity in the forest. The principal behind boosting, on the other hand, is to break the independence assumption, instead using the result of one model to guide the training of the next. An immediate consequence of this is that random forests can be parallelized, whereas boosted forests need be trained in sequence.

Algorithm 12.6: AdaBoost algorithm for classification.

Input: data $\mathcal{D}$;
feature set $\mathcal{F}$;
weighted learning algorithm $\mathscr{A} : \mathbb{D} \times \mathbb{F} \times \mathbb{R}^n_+ \to \mathbb{M}$;
the size m of the ensemble
Output: An ensemble model $\mathcal{E}$.

1 $w_i = 1/|\mathcal{D}|$, for $i = 1, \ldots, |\mathcal{D}|$
2 **for** j **from** 1 **to** m **do**
3 $\quad$ Train model using weights $\mathbf{w} = \langle w_1, \ldots, w_n \rangle$: $M_j = \mathscr{A}(\mathcal{D}, \mathcal{F}, \mathbf{w})$
4 $\quad$ Compute weighted training error: $\epsilon = \dfrac{\sum\limits_{i=1}^{n} w_i \mathbb{I}[y_i \neq M_j(x_i)]}{\sum_{i=1}^{n} w_i} < 1/2$
5 $\quad$ Compute the accuracy log odds: $\alpha_j = \log\left(\dfrac{1-\epsilon}{\epsilon}\right)$
6 $\quad$ Update weights: $w_i = w_i \exp\left(\alpha_j \mathbb{I}[y_i \neq M_j(x_i)]\right)$
7 **end**
8 **return** $\mathcal{E} = \{M_j, \alpha_j\}_{j=1}^m$

The most outstanding difference between our first boosting method, Algorithm 12.6, and bagged forests, Algorithm 12.5, is that the base learner $\mathscr{A}$ requires the ability to handle *weights* to the training set. Fortunately, we have already taken care to include the handling of training weights in our decision tree implementation of Code Blocks 12.4–12.6, so that we can use our python DecisionTree class straightaway.

In Algorithm 12.6, the weights are initialized to uniform $(1/|\mathcal{D}|)$, but updated during each step of training (line 6). The weights are used to train each model on line 3. Upon training each tree, however, we then calculate the weighted training error ϵ, and an associated log odds, defined on line 5. Since we assume that each base model is a weak learner, we are assuming that the training error is $\epsilon < 1/2$, which, in turn, ensures that the log odds α_i for iteration i is positive, $\alpha_i > 0$. Finally, notice that only the weights for the *misclassified* instances of model M_i are updated by a factor of $e^{\alpha_i} > 1$; correctly classified weights remain the same. This has the effect of amplifying the weights of the misclassified examples, or errors, giving them more importance for the next round of model training. In this way, each model pays greater attention to the instances corresponding to the mistakes of the past.

The Python implementation of Algorithm 12.6 is shown in Code Block 12.18. Since we constructed our DecisionTree class to handle weighted predictions, we only need overwrite the _trainNext method of the Ensemble abstract base class.

```python
class AdaBoost(Ensemble):

    def _getModel(self):
        return DecisionTree

    def _trainNext(self, df, y):
        if not hasattr(self, 'boost_weights'):
            self.boost_weights = np.ones(len(df)) / len(df)
        model = self.Model(**self.model_params)
        model.train(df, y, weights=self.boost_weights)
        y_hat = model.predict(df)
        err = np.average(y_hat!=y, weights=self.boost_weights)
        alpha = np.log((1-err) / err)
        self.boost_weights *= np.exp(alpha * (y_hat != y))
        return model, alpha
```

Code Block 12.18: AdaBoost implementation.

12.3.2 Gradient Boosting

Gradient boosting was first introduced in Friedman [2001], and updated to include sampling (stochastic gradient boosting) in Friedman [2002]. We will begin with an overview of gradient boosting, primarily focused on regression, before discussing regression and classification in detail.

Gradient Boosting Overview

The basic structure is shown in Algorithm 12.7. Like AdaBoost, the gradient boosting algorithms makes successive corrections to the previous model, so that each model improves on the prior model's weakness. In the case of gradient boosting, however, this is done relative to a particular loss function, which is used as an objective function that the ensemble seeks to minimize. Note that since the models in the gradient boosting ensemble are meant as literally corrections to each other, we require an unnormalized ensemble.

Naturally, the gradient boosting Algorithm 12.7 is a generalization of the steepest descent Algorithm 5.2, and the stochastic gradient boosting

---

Algorithm 12.7: GradientBoostedRegressor algorithm for gradient boosting ($f = 1$) or stochastic gradient boosting ($f \in (0,1)$) for regression.

Input: data $\mathcal{D}$;
      feature set $\mathcal{F}$;
      regression learning algorithm $\mathscr{A} : \mathbb{D} \times \mathbb{F} \to \mathbb{M}$;
      a differentiable loss function $L : \mathbb{T} \times \mathbb{T} \to \mathbb{R}$;
      the size m of the ensemble;
      subsample fraction $f \in (0,1]$
Output: An ensemble model $\mathcal{E}$.

1 Initialize model to a constant: $M_0(x) = \arg\min_{\gamma} \sum_{i=1}^{n} L(y_i, \gamma)$

2 Initialize an unnormalized ensemble $\mathcal{E} = \{M_0, 1\}$

3 **for** j from 1 to m **do**

4   Compute sample index $I_{\mathcal{B}} = \texttt{getRandomIndex}(|\mathcal{D}|, \texttt{int}(f|\mathcal{D}|))$

5   Compute the gradients: $r_i = -\left. \dfrac{\partial L(y_i, \gamma)}{\partial \gamma} \right|_{\gamma = \mathcal{E}(x_i)}$, for $i \in I_{\mathcal{B}}$

6   Train a *gradient model*: $M_j = \mathscr{A}\left(\{(x_i, r_i)\}_{i \in I_{\mathcal{B}}}, \mathcal{F}\right)$

7   Solve the one-dimensional optimization problem:

$$\gamma_j = \arg\min_{\gamma > 0} \sum_{i=1}^{n} L(y_i, \mathcal{E}(x_i) + \gamma M_j(x_i))$$

8   Update $\mathcal{E} = \mathcal{E} \oplus \{M_j, \gamma_j\}$

9 **end**

10 **return** $\mathcal{E}$

---

is based on the stochastic gradient descent variation, as defined in Definition 5.5. Let's first discuss the stochastic nature of the algorithm, and then dive in to the details of the gradient boosting itself.

The update from gradient boosting to stochastic gradient boosting is achieved with the *subsample fraction* parameter, which, for regular gradient boosting, has the value $f = 1$. We then add the `getRandomIndex(n,m)` function (line 4), which returns a random selection of m indices from the the full index array $\{1, \ldots, n\}$; i.e., it is a random selection of indices for a bootstrap without replacement. For $f = 1$, we have $I_B = \{1, \ldots, n\}$, so that the full data set is used in each iteration of the algorithm. A typical value of the subsample fraction is $f = 0.5$, though it should be viewed as a tunable parameter.

Now that we understand how to extend the gradient boosting algorithm to its stochastic counterpart, let's next dive into the details of the algorithm itself. The first departure from our prior methods is that gradient boosting requires a loss function, which it then seeks to optimize for. We defined loss functions in Equation (6.18), and defined common examples in Equations (6.17)–(6.20). Typically, we choose squared-error loss (Equation (6.17)) for regression and log loss (Equation (6.20)) for classification, though the latter does require the predicted class probabilities, not the predicted labels.

We repeat the squared-error loss and log loss functions for convenience:

$$L(y, \gamma) = (y - \gamma)^2, \tag{12.14}$$

$$L(y, (p_1, \ldots, p_c)) = -\sum_{k=1}^{c} \mathbb{I}[y = C_k] \log p_k. \tag{12.15}$$

The log loss is also often referred to as the *deviance*, due to its relation to the deviance function (Definition 7.12) from logistic regression. Moreover, note that the negative gradient of the squared-error loss is simply

$$-\frac{\partial L}{\partial \gamma} = 2(y - \gamma),$$

which is simply twice the residual error, when $\gamma = \hat{y}$ is interpreted as the model prediction. This gives a further interpretation of gradient boosting in the context of regression, where the models learned on line 6 are trying to learn the *residual errors* of the previous ensemble.

Given a loss function, the algorithm is then a simple generalization of gradient descent (Algorithm 5.2), such that steps in the direction of the negative gradient are instead replaced with steps in the direction of a model trained on the negative gradient. The optimization problem on line 7 implies that our gradient boosting can be viewed as a steepest-descent gradient boosting algorithm. In some instances, the parameter γ is passed as an input, called the *learning rate*, and is held constant over the iterations. The stochastic variation follows in the same suit as the stochastic gradient descent algorithm of Definition 5.5, where only a subsample of data is used for each step.

Finally, we stress the importance of using an *unnormalized* ensemble in the construction of the ensemble $\mathcal{E}$. If we instead denote the model trained on line 6 by R_j (for residual), we could then replace line 8 with

$$M_j = M_{j-1} + \gamma_j R_j,$$

to the same effect. The final ensemble should therefore necessarily generate predictions of the form

$$\mathcal{E}(x) = M_0(x) + \gamma_1 M_1(x) + \cdots + \gamma_m M_m(x),$$

as $M_0(x)$ is an overall (constant) prediction and as each subsequent model $M_1, \ldots, M_m$ is a corrective term seeking to remedy the errors of the prior models. It would therefore be incorrect to divide this final prediction by the normalization factor $1 + \sum_{j=1}^{m} \gamma_j$.

Gradient Boosting for Regression

We begin by noting that squared-error loss Equation (12.14) directly yields the MSR impurity function defined in Equation (12.9), when summed over a sample of data. Next, we note that, when using squared-error loss, line 1 of Algorithm 12.7 yields the sample mean:

$$\arg\min_{\gamma} \sum_{i=1}^{n} (y_i - \gamma)^2 = \overline{y}. \tag{12.16}$$

We encoded a quick `ConstantModel` class, which outputs this constant, in Code Block 12.19. The full gradient boosting algorithm for regression is encoded in Code Block 12.20.

Gradient Boosting for Classification: Initial Model

As we saw with regression, log loss, as defined by Equation (12.15), yields the entropy measure of Equation (12.4), when summed over a sample of data, and when we identify p_k as the sample mean for the kth class. More directly, we obtain the following:

$$\mathscr{L}(p; \mathcal{D}) = \sum_{i=1}^{n} L(y_i, p) = -\sum_{k=1}^{c} \overline{y}_k \log_2(p_k), \tag{12.17}$$

where $y_k = \mathbb{I}[y_i = \mathcal{C}_k]$, so that

$$\overline{y}_k = \frac{1}{n} \sum_{i=1}^{n} \mathbb{I}[y_i = \mathcal{C}_k]$$

```
 1  class ConstantModel(Model):
 2
 3      def __init__(self, **kwargs):
 4          type_error = "Keyword argument 'type' must be set to
                  classification or regression"
 5          assert 'type' in kwargs, type_error
 6          assert kwargs['type'] in ['classification', 'regression'],
                  type_error
 7          self.type = kwargs['type']
 8
 9      def train(self, df, y, weights=None):
10          if self.type == 'regression':
11              self.constant = np.average(y, weights=weights)
12          else:
13              self.constant = Series(Counter(y)).sort_index().values /
                      len(y)
14
15      def predict(self, df: DataFrame):
16          return self.constant
```

Code Block 12.19: `ConstantModel` class representing a constant prediction.

is the sample average. The values of the vector p in Equation (12.17) are not free to vary at will, but, rather, are constrained by the condition that

$$p \in \Delta^{c-1} = \left\{ p \in \mathbb{R}^c_* : g(p) = \sum_{k=1}^c p_k = 1 \right\};$$

i.e., p is an element of the probability simplex (Definition 2.23). The arg min of Algorithm 12.7 line 1, therefore, must be modified as a constrained argmin:

$$M_0(x) = \arg\min_{p \in \Delta^{c-1}} \mathcal{L}(p; \mathcal{D}).$$

To solve this, we write down the Lagrange multiplier equations

$$\nabla \mathcal{L} = \lambda \nabla g,$$

$$g(p) = \sum_{k=1}^c p_k = 1,$$

from whence we obtain

$$-\frac{\overline{y}_k}{p_k} = \lambda,$$

so that $p_k = -\lambda^{-1}\overline{y}_k$. The constraint equation implies that $\lambda = -1$, since $\sum_{k=1}^c \overline{y}_k = 1$. This yields the result that we should initialize our probability

```
1   class GradientBoostedRegressor(Ensemble):
2
3       def __init__(self, **kwargs):
4           kwargs['normalize'] = False
5           kwargs['type'] = 'regression'
6           self.subsample = kwargs.get('subsample', 1)
7           super().__init__(**kwargs)
8
9       def _getModel(self):
10          return DecisionTree
11
12      def _trainNext(self, df, y):
13          if len(self.models) == 0:
14              M = ConstantModel(type=self.type)
15              M.train(df, y)
16              return M, 1
17          size = int(self.subscample * len(df))
18          idx = np.random.choice(np.arange(len(df)), size=size,
                  replace=False)
19          df_boot = df.loc[idx, :].reset_index(drop=True)
20          y_boot = y[idx]
21          y_hat = self.predict(df_boot)
22          r = y_boot - y_hat
23          model = self.Model(**self.model_params)
24          model.train(df_boot, r)
25          gamma = goldenSearch(lambda x: mean_squared_error(y,
                  self.predict(df) + x*model.predict(df)), 0, 10)
26          return model, gamma
```

Code Block 12.20: `GradientBoostedRegressor` gradient boosted regressor class.

model with the constant predictions coinciding with the observed sample ratios from each class:

$$M_0(x) = \frac{1}{n} \left\langle \sum_{i=1}^{n} \mathbb{I}[y_i = C_1], \dots, \sum_{i=1}^{n} \mathbb{I}[y_i = C_c] \right\rangle = \langle \bar{y}_1, \dots, \bar{y}_c \rangle. \quad (12.18)$$

This initial model coincides with the observed probability prediction in the `ConstantModel` class of Code Block 12.19.

Gradient Boosting for Classification: Boosting Rounds

In many machine-learning texts, the gradient boosting classification algorithm is handled by training c independent binary classifiers for each of the c classes, so that the for-loop of lines 3–9 in Algorithm 12.7 is actually nested within an outer for-loop that iterates through the target labels (classes). We will present an alternate approach, which we believe to be more elegant, that treats the probability p as a vector $p \in \Delta^{c-1}$ and that addresses the geometric constraints directly within the algorithm.

Our gradient-boosted classification algorithm is given by Algorithm 12.8. In order to unravel this, we begin by introducing the normal vector to the constraint $g(p) = \sum_{k=1}^{c} p_k = 1$,

$$v = \nabla g = \langle 1, \dots, 1 \rangle \in \mathbb{R}^c, \tag{12.19}$$

with norm $\|v\|_2 = \sqrt{c}$. The tangent plane $T_p \Delta^{c-1}$ to a point $p \in \Delta^{c-1}$ is therefore isomorphic to the orthogonal complement of the subspace spanned by the vector v. In order to modify Algorithm 12.7 for classification, we therefore need only project the residuals, as defined on line 5 of the algorithm, onto the tangent plane of the constraint surface, using the projection

---

Algorithm 12.8: GradientBoostedClassifier algorithm for gradient boosting ($f = 1$) or stochastic gradient boosting ($f \in (0,1)$) for classification; using log loss (Equation (12.15)).

Input: data $\mathcal{D}$;
     feature set $\mathcal{F}$;
     a vector regression learning algorithm $\mathscr{A} : \mathbb{D} \times \mathbb{F} \to \mathbb{M}$;
     the size m of the ensemble;
     subsample fraction $f \in (0, 1]$
Output: An ensemble model $\mathcal{E}$.

1   Initialize model to a constant:

$$M_0(x) = \frac{1}{n} \left\langle \sum_{i=1}^{n} \mathbb{I}[y_i = C_1], \dots, \sum_{i=1}^{n} \mathbb{I}[y_i = C_c] \right\rangle$$

2   Initialize an unnormalized ensemble $\mathcal{E} = \{M_0, 1\}$

3   **for** j from 1 to m **do**

4   Compute sample index $I_\mathcal{B} = \texttt{getRandomIndex}(|\mathcal{D}|, \texttt{int}(f|\mathcal{D}|))$

5   Compute the gradients: $r_i = \left\langle \dfrac{\mathbb{I}[y_i = C_1]}{\mathcal{E}(x_i)_1}, \dots, \dfrac{\mathbb{I}[y_i = C_c]}{\mathcal{E}(x_i)_c} \right\rangle$, for $i \in I_\mathcal{B}$

6   Compute the projections: $v_i = \text{proj}_{T_{\mathcal{E}(x_i)} \Delta^{c-1}}(r_i)$, for $i \in I_\mathcal{B}$

7   Train a *gradient model*: $M_j = \mathscr{A}(\{(x_i, v_i)\}_{i \in I_\mathcal{B}}, \mathcal{F})$

8   Compute $\gamma^{\max} = \min\limits_{i=1,\dots,n} \min\limits_{k=1,\dots,c} \left\{ \dfrac{-\mathcal{E}(x_i)_k}{M_j(x_i)_k} : M_j(x_i)_k < 0 \right\}$

9   Solve the one-dimensional optimization problem:

$$\gamma_j = \operatorname*{arg\,min}_{\gamma \in (0, 0.9\gamma^{\max})} \sum_{i=1}^{n} L(y_i, \mathcal{E}(x_i) + \gamma M_j(x_i))$$

10   Update $\mathcal{E} = \mathcal{E} \oplus \{M_j, \gamma_j\}$

11   **end**

12   **return** $\mathcal{E}$

---

operator

$$\text{proj}_{T_p \Delta^{c-1}}(x) = x - \frac{x \cdot v}{c} v, \tag{12.20}$$

which maps a vector $x \in \mathbb{R}^c$ onto the tangent plane $T_p \Delta^{c-1} \cong \text{span}(v)^{\perp}$.

Next, we need to consider the boundary of the simplex Δ^{c-1}, for the purpose of further constraining the optimization problem defined on line 7 of Algorithm 12.7. Given a point $(p, v) \in T\Delta^{c-1}$ on the tangent bundle (i.e., a probability vector and an attached tangent vector), we need to determine the shortest distance to the edge of the simplex, in the direction $v \in T_p \Delta^{c-1}$. By traveling a distance γ in the direction v, we reach the kth coordinate plane precisely when

$$p_k + \gamma_k v_k = 0, \qquad \text{or}, \qquad \gamma_k = -\frac{p_k}{v_k}.$$

Considering only the directions for which $\gamma_k > 0$ (which only occurs when the vector v has a negative value for its kth component), we then simply select the minimum parametric distance

$$\gamma = \min_{k=1,\ldots,c} \left\{ -\frac{p_k}{v_k} : v_k < 0 \right\}.$$

All modifications in hand, the final gradient-boosted classification algorithm is given in Algorithm 12.8.

The Python code for the gradient-boosted classifier is given in Code Blocks 12.20 and 12.21. We immediately encounter two differences with our other forest methods right away: the _logLoss and predictProba methods on lines 12 and 18 of Code Block 12.20. The log-loss function (line 12) is specifically written for a vector y of class labels and a probability matrix p, of the same length as the vector y and a number of columns equal to the total number of classes.

For line 21 of the predictProba method of Code Block 12.20, note that the first model (self.models[0]) will be an object from the ConstantModel class (Code Block 12.19), whereas subsequent models will be regression DecisionTree objects. (Note that our original DecisionTree class, of Code Blocks 12.3–12.6, was constructed with the ability to perform *vector regression*, where the target variable is a vector variable. This is why, for example, we specify axis=0 in the _mse method on line 19 of Code Block 12.4, etc.)

12.4 Advanced Forestry

In this section, we will briefly discuss some additional variations of the classic random forests and gradient boosted forests. We begin by mentioning a powerful variation of gradient boosting known as extreme gradient boosting. We will call out the main differences and point the interested reader to

```
 1   class GradientBoostedClassifier(Ensemble):
 2
 3       def __init__(self, **kwargs):
 4           kwargs['normalize'] = False
 5           kwargs['type'] = 'classification'
 6           self.subsample = kwargs.get('subsample', 1)
 7           super().__init__(**kwargs)
 8
 9       def _getModel(self):
10           return DecisionTree
11
12       def _logLoss(self, y, p):
13           logp = np.log2(p)
14           y_labels = [self.classes.index(x) for x in y]
15           logp = [logp[i, y_labels[i]] for i in range(len(y))]
16           return -np.sum(logp)
17
18       def predictProba(self, df):
19           y_hat = np.zeros((len(df), len(self.classes)))
20           for model, weight in zip(self.models, self.weights):
21               y_hat += weight * model.predict(df)
22           return y_hat
23       # continued ...
```

Code Block 12.21: `GradientBoostedClassifier` gradient boosted classifier class (part 1).

resources for this variation. Next, we will discuss two variations that can be used for temporal data. The first is survival forests, which is a modification of the random forest algorithm to account for censored information. The second, we will call online forests.

12.4.1 Parallel Processing

As modern laptops come with a minimum of eight CPUs, it is worth mentioning how we can speed up the training of forests by leveraging parallel processing. Python comes with a built-in module `multiprocessing` that can be used to parallelize a job ofer multiple processors. Unfortunately, there is no easy way to code parallel processing as a method within a class, as the `multiprocessing` module does not work interactively in the Python interpreter. (We will discuss an alternate approach at the end of the section.) Instead, we must write an executable script that we can run. The idea, then, is to train a number of smaller forests in parallel, and add them together to get our end result. We therefore begin by defining the magic method `__add__` to our `Ensemble` class (Code Block 12.9), which allows us

```
1   # ... continued
2   def _trainNext(self, df, y):
3       if len(self.models) == 0:
4           M = ConstantModel(type='classification')
5           M.train(df, y)
6           return M, 1
7       size = int(self.subsample * len(df))
8       idx = np.random.choice(np.arange(len(df)), size=size,
                replace=False)
9       df_boot = df.loc[idx, :].reset_index(drop=True)
10      y_boot = y[idx]
11      y_hat = self.predictProba(df)
12      y_hat_boot = y_hat[idx]
13      r = [[int(y==c) for c in self.classes] for y in y_boot] /
                y_hat_boot
14      r = r - r.sum(axis=1).reshape((len(r), 1)) /
                len(self.classes)
15      self.model_params['type'] = 'regression'
16      model = self.Model(**self.model_params)
17      model.train(df_boot, r)
18      v = model.predict(df)
19      gamma = - y_hat / v
20      gamma_max = min(gamma[gamma > 0])
21      gamma = goldenSearch(lambda x: self._logLoss(y,
                self.predictProba(df) + x*model.predict(df)), 0,
                0.9*gamma_max)
22      return model, gamma
23
24  def predict(self, df):
25      y_hat = self.predictProba(df)
26      votes = DataFrame(y_hat, columns=self.classes)
27      return votes.idxmax(axis=1).values.tolist()
```

Code Block 12.22: `GradientBoostedClassifier` gradient boosted classifier
class (part 2).

to use the syntax `E + F`, which is interpreted by Python as `E.__add__(F)`.
The code to accomplish this is shown in Code Block 12.23.

Finally, we can use the `Pool` function from the `multiprocessing` module to parallelize our computation, as shown in Code Block 12.24. Here, we
use `starmap`, as our function takes multiple arguments. We ran this on a
10-CPU machine, so that each CPU trained four separate random forests,
yielding a total of 400 individual decision trees that were combined in the
end.

```
1   def __add__(self, other):
2       assert self.type == other.type
3       assert self.Model == other.Model
4       new = copy.deepcopy(self)
5       w_new, w_other = sum(new.weights), sum(other.weights)
6       new.models += other.models
7       new.weights += other.weights
8       if hasattr(self, 'oob_score'):
9           new.oob_score = w_new * new.oob_score + w_other *
                other.oob_score / (w_new + w_other)
10      return new
```

Code Block 12.23: Magic method for Ensemble class (Code Block 12.9) for adding forests.

```
1   def fun(E, df, y):
2       E.train(df, y)
3       return E
4
5   def main():
6       data = load_diabetes()
7       X = data.data
8       y = data.target
9       df = DataFrame(X, columns = data.feature_names)
10      df.sex = df.sex.apply(lambda x: 'M' if x > 0 else 'F')
11      num_features = ['age', 'bmi', 'bp', 's1', 's2', 's3', 's4',
            's5', 's6']
12      E = RandomForest(type='regression', cat_features=['sex'],
13                  num_features=num_features,
14                  max_depth=3, min_samples_split=20,
15                  subspace_dim=3, size=10)
16      start = time.time()
17      values = ((E, df, y) for i in range(40))
18      with multiprocessing.Pool(processes=10) as pool:
19          res = pool.starmap(fun, values)
20      E = res[0]
21      for i in range(1, len(res)):
22          E = E + res[i]
23      end = time.time()
24      print(f'elapsed time: {end - start}')
25
26  if __name__ == '__main__':
27      main()
```

Code Block 12.24: Using multiprocessing to train trees in parallel.

Accessing the Model Object

When running a script, such as Code Block 12.24, the variables introduced within the **main** function are not retained in memory. Once the script completes, the local memory is cleared. In a development platform, such as Spyder, one is often interested in running a script, but retaining the final model output in memory, so that it may be accessed from the iPython console for further investigation. There are two approaches that allow one to persist the model constructed within **main** in the runtime environment.

Return the Model Object

The most straightforward approach is simply to add a line **return E** at the end of the **main** function, and then modify the last line of Code Block 12.24 to store the return in memory: **E = main()**. When running the script in the iPython console, such an arrangement will then persist the final model object **E** in memory.

Pickling Model Objects

Another approach, which can be useful both in development platforms as well as production platforms, is pickling the model. *Pickling*[3] refers to a process that takes an object and converts it into an equivalent text representation, often but not necessarily binary, that retains all of the information within the object hierarchy. A copy of the object can then be reconstructed by *unpickling* the pickle file that was rendered. Pickling can be useful in production, as it allows one to save a trained model on the server or on a distributed file system, such as HDFS, for later use.

In order to pickle and unpickle an object, one simply imports the **pickle** module and applies the **dump** and **load** functions, respectively, which pickle the object and save the result to disk, as shown in lines 3–7 of Code Block 12.25[4]. One can add lines 3 and 4 to the **main** function, and then call lines 6 and 7 from the iPython console to retrieve the model object **E** created during execution of the script. Alternatively, one may use **dumps** and **loads** to convert the object to a string and save to local memory, as shown in lines 9 and 10 of Code Block 12.25.

Parallel Processing with `joblib`

An alternate approach to parallel processing is the `joblib` module, which has the advantage of being usable within the iPython console or within a given class method. For our use case, we can build the parallel processing

---

[3] also referred to as serialization or marshalling.

[4] The 'b', as in 'wb' (write) or 'rb' (read), instructs Python to write to and read the file in binary.

```python
import pickle
# pickle object (e.g., from main function)
with open('tmp_files/my_model.pkl', 'wb') as f:
    pickle.dump(E, f)
# unpickle object (e.g., from iPython console)
with open('tmp_files/my_model.pkl', 'rb') as f:
    E = pickle.load(f)
# pickle and unpickle object in memory
pkl = pickle.dumps(E)
E_new = pickle.loads(pkl)
```

Code Block 12.25: Pickling and unpickling objects.

functionality directly into the **train** method of the **Ensemble** class (Code Block 12.9), as shown in Code Block 12.26. Note that the optional keyword argument **n_jobs** was provided in the constructor of the **BaggedForest** class (Code Block 12.14). If **self.n_jobs** is undefined, or is equal to its default value of 1, we preserve the original code from Code Block 12.9. Otherwise, we use **Parallel** and **delayed** to run the individual calls to **self._trainNext** in parallel.

```python
from joblib import Parallel, delayed
## train method
        if not hasattr(self, 'n_jobs') or self.n_jobs == 1:
            for i in tqdm(range(self.size)):
                model, weight = self._trainNext(df, y)
                self.models.append(model)
                self.weights.append(weight)
        else:
            results = 
                Parallel(n_jobs=self.n_jobs)(delayed(self._trainNext)(df,
                y) for i in tqdm(range(self.size)))
            self.models, self.weights = map(list, zip(*results))
```

Code Block 12.26: Modification of the **Ensemble** class (Code Block 12.9) for parallel processing.

Note 12.1. Recall that parallel processing is not appropriate for boosted forests, as boosted forests algorithms are defined sequentially. For this reason, we added the definition of **self.n_jobs** to the constructor of the **BaggedForest** class (Code Block 12.14), so that the parallel processing cannot be used for non-bagged descendants of the **Ensemble** class, where they are bound to cause side effects or not function properly.   ▷

12.4.2 Extreme Gradient Boosting

Extreme gradient-boosted trees, or *XGBoost*, is an open-source software (`xgboost.readthedocs.io`) that represents a powerful extension to simple gradient boosting; see Chen and Guestrin [2016]. Not only does XG-Boost improve upon the basic gradient-boosted forest methods, but it also improves processing performance, optimizing for training speed. Improvements on the algorithm include:

1. The Newton–Raphson method (Algorithm 5.1) replaces gradient descent, allowing for more efficient optimization, using the Hessian matrix[5] of the loss function;
2. a regularization term (Lasso L1 or Ridge L2; Definition 7.4) is added to the loss function to penalize complex models and avoid overfitting;
3. a *weighted quantile sketch* algorithm is used to efficiently find optimal splits;
4. built-in cross validation at each iteration;
5. an added randomization parameter that helps reduce correlation between trees;
6. a method for proportionally "shrinking" leaf nodes;

In addition, the algorithm provides various improvements for processing performance:

1. the for-loops for calculating features are parallelized using multi-threading, improving computational speed;
2. tree pruning uses a 'depth-first' approach to improve computational speed;
3. the algorithm optimizes hardware memory usage using 'cache awareness,' to optimally allocate memory of the gradient within each thread, and 'out-of-core' computing, to optimize disk space when training large datasets that do not fit into memory.

An `XGBoost` class is provided in Code Block 12.27. This is essentially a wrapper around the `xgboost` module that reformats the inputs and outputs to follow our conventions, thereby making it compatible and exchangeable with our other methods.

12.4.3 Bagged Boosters

One of the limitations of gradient boosting is its sequential nature, which at first glance prevents use of parallel computations. Recall, however, that bagged forests are simply bootstrapped aggregations of individual models. There is no reason why we therefore cannot do a boostrap aggregation using a gradient descent model as the kernel. In this regard, we are trading some of our depth, as measured by the number of boosting rounds, with breadth,

---

[5] matrix of second partial derivatives

```
1   import xgboost
2   class XGBoost(Model):
3
4       def __init__(self, **kwargs):
5           # See https://xgboost.readthedocs.io/en/stable/parameter.html
6           self.objective = kwargs.get('objective', 'reg:squarederror')
7           self.size = kwargs.get('size', 10)
8           keys = ['eta', 'gamma', 'max_depth', 'min_child_weight',
                    'subsample', 'sampling_method', 'colsample_bytree',
                    'alpha', 'lambda', 'tree_method', 'scale_pos_weight',
                    'max_leaves', 'num_parallel_tree']
9           self.params = {key: kwargs[key] for key in
                    set(keys).intersection(kwargs.keys())}
10
11      def train(self, df, y, weights=None):
12          dmat = xgboost.DMatrix(df, label=y, enable_categorical=True)
13          self.xgb = xgboost.train(self.params, dmat, self.size)
14
15      def predict(self, df, get_prob=False):
16          dmat = xgboost.DMatrix(df, enable_categorical=True)
17          return self.xgb.predict(dmat)
```

Code Block 12.27: Wrapper around XGBoost package.

as measured by the number of boosted models in the forest. The advantage is that we can easily parallelize the individual boosters. This is achieved in Code Block 12.28, which provides code for a random gradient-boosted forest and a random XGBoosted forest.

```
1   class RandomGBForest(BaggedForest):
2
3       def _getModel(self):
4           self.model_params['surpress_tqdm'] = True
5           if self.type == 'classification':
6               return GradientBoostedClassifier
7           return GradientBoostedRegressor
8
9   class RandomXGBForest(BaggedForest):
10
11      def _getModel(self):
12          return XGBoost
```

Code Block 12.28: Random Gradient Boosted and Extreme Gradient Boosted Forests.

12.4.4 Performance Comparison

Next, let's compare the performance—in terms of both speed and model accuracy—of the various tree-based methods presented thus far. We use the *sci-kit learn* diabetes dataset to train a number of regression models, as shown in Code Block 12.29. In particular, we train the set of models contained in the dictionary that spans lines 10–18. The results are shown in Figure 12.7.

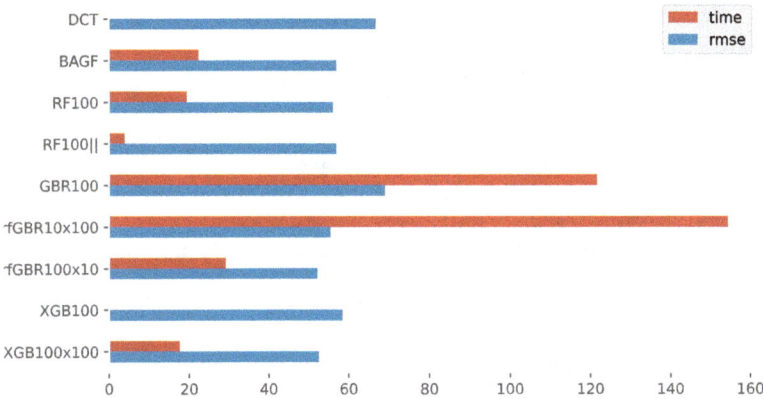

Fig. 12.7: Speed and performance comparison on a sample data set.

The first observation is that there is very little variance in the model's error (blue), so each model does a good job learning from the data. Naturally, a single decision tree (DCT) was quite fast, clocking in around 1/5th of a second. It also had the worst error, though fairly comparable to a single gradient-boosted forest (GBR100), which took around 120s to train. (It is quite surprising the gradient-boosted regressor offered little performance improvement over a single decision tree.)

The bagged forest (BAGF) and random forest (RF100) performed a bit better than a single decision tree, as expected, and the random forest was a tad faster than the bagged, due to the smaller feature space. Running the random forest in parallel (RF100‖) further improved speed by a factor of five.

We trained two random gradient-boosted forests: the first consisted of 10 gradient boosted forests with 100 trees each (rfGBR10x100), and the second consisted of 100 gradient boosted forests with 10 trees each (rfGBR100x10). The first only took around 30% longer than a single 100-tree forest (GBR100) due to the parallelization, whereas the second was

```
1   def main():
2       data = load_diabetes()
3       X = data.data
4       y = data.target
5       df = DataFrame(X, columns = data.feature_names)
6       df.sex = df.sex.apply(lambda x: 'M' if x > 0 else 'F')
7       df.sex = df.sex.astype('category')
8       num_features = ['age', 'bmi', 'bp', 's1', 's2', 's3', 's4',
            's5', 's6']
9       params = {'type':'regression', 'cat_features':['sex'],
            'num_features':num_features, 'max_depth':3,
            'min_samples_split':20, 'min_child_weight':20,
            'subspace_dim':3, 'subsample':0.5}
10      models = {'DCT': DecisionTree(**params),
11      'BAGF': BaggedForest(**params, size=100),
12      'RF100': RandomForest(**params, size=100),
13      'RF100||': RandomForest(**params, size=100, n_jobs=10),
14      'GBR100': GradientBoostedRegressor(**params, size=100),
15      'rfGBR10x100': RandomGBForest(**params, size=10, n_jobs=10,
            model_params={'size':100}),
16      'rfGBR100x10': RandomGBForest(**params, size=100, n_jobs=10,
            model_params={'size':10}),
17      'XGB100': XGBoost(**params, size=100,
            objective='reg:squarederror'),
18      'rfXGB100x100': RandomXGBForest(**params, size=100,
            objective='reg:squarederror')}
19      dr = DataFrame(columns=['model', 'time', 'rmse'])
20      df_train, df_test, y_train, y_test = train_test_split(df, y,
            test_size=0.2)
21      df_train, df_test = df_train.reset_index(drop=True),
            df_test.reset_index(drop=True)
22      for model, E in models.items():
23          start = time.time()
24          E.train(df_train, y_train)
25          t = time.time() - start
26          y_hat = E.predict(df_test)
27          mse = mean_squared_error(y_test, y_hat)
28          dr = dr.append(DataFrame({'model':model, 'time':t,
                'rmse':np.sqrt(mse)}, index=[0]), ignore_index=True)
29          print(f"model: {model}, time: {t}, mse: {mse}")
30      return dr
31
32  if __name__ == '__main__':
33      dr = main()
34      dr = dr.set_index('model')
35      plt.figure(figsize=(8,9/2))
36      plt.style.use('ggplot')
37      ax = dr.plot.barh(figsize=(8,9/2))
38      ax.invert_yaxis()
```

Code Block 12.29: Model comparison.

80% faster than the single 100-tree forest, with better performance. Both rfGBR10x100 and rfGBR100x10 consist of 1,000 trees, but the later had superior performance in terms of both speed and accuracy.

The XGBoosted forest (XGB100) with 100 boosted rounds was as fast as our implementation of a single decision tree! The XGBoost package, however, is optimized to use parallelization and multi-threading, though it is interesting that it is around 100 times faster than our random forest implementation. Finally, we trained a bagged XGBoosted forest consisting of 100 individual XGBoosted forests with 100 boosted rounds. This offered an additional, albeit modest, reduction in the error. Interestingly, it made no difference in terms of speed whether we included the n_jobs=10 parameter, leading us to infer that the xgboost package is already leveraging the multi-processing capacity of the machine.

Finally, though not shown in the figure, we note that the *sci-kit learn* models RandomForestRegressor and GradientBoostingRegressor, from the sklearn.ensemble package, both performed quite well, clocking in around 1/7th and 1/20th of a second, respectively, and achieving a RMSE of around 56.

12.5 Time Forests and Survival Forests and the Like

Time plays an important role in many real-world applications. This slightly breaks from the view of traditional machine learning, in which one trains a model over a static data set; e.g., one might train a classifier over a library of pictures to teach the machine how to classify cats and dogs. Many, if not most, interesting data science applications in marketing and product analytics have an explicit time component: companies collect massive quantities of user behavioral data over time. It is therefore critical to adapt traditional machine-learning models to settings that have such a temporal focus. We discuss time aspects in two contexts: the first is how to handle time as a dimension, and the second is how to model online and survival processes (Definition 5.20). This latter problem is trickier, as there are actually two time dimensions, the common case being a time series of right-censored data, forming a sort of *lower-right-triangular censoring* (e.g., see Figure 5.10).

12.5.1 Time Forests

We begin with the topic of sequential data sets indexed by time. For example, a marketing campaign might run over a long period of time, each day generating a number of impressions and clicks. If a set of user-level features is available at the impression level, one might model the probability of a click as a function of each user's feature set. A large-scale campaign might generate hundreds of millions of impressions each day. Instead of training a single model using the trailing 30-days of data, and repeating this process each day, one might instead train a differential model, only on the prior day's cohort, and maintain a running queue of trailing models. This is the idea behind what we call *time forests*.

Note 12.2. Mathematically, time is not topologically different than a simple index. In practice, however, *time* sets the temporal cadence for which we receive and can process data. ▷

Definition 12.12. *A* time forest *of size m and* decay factor γ *is a normalized ensemble, such that individual models and weights are stored in a queue with maximum length m, and such that the ensemble's* train *operation*

1. *applies the decay factor γ to all weights,*
2. *trains its base model on the new input data, generating a new model M,*
3. *if the model and weight queue is length m, pops the oldest model and weight from the left (beginning) of the queue,*
4. *appends the new model M and its associated weight (i.e., size of data set) to the right of the queue.*

A time forest of size 7 is illustrated in Figure 12.8. The *pop* operation corresponds to removing the leftmost model and its associated weight. The *append* operation appends a new model and weight to the right. The oldest model is the leftmost model, as shown by the model dates.

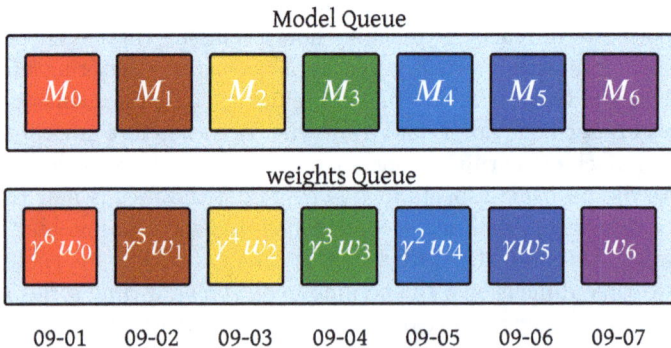

Fig. 12.8: Illustration of a *time forest* of size 7 consisting of a model queue and a weights queue.

Note 12.3. Often, the base model of a time forest is itself a forest, e.g., a random forest or an XGBoosted forest. Like our *bagged boosters*, time forests also represent a *forest of forests*; i.e., ensembles built over forest base models. ▷

Note 12.4. The weights normalization in the `Ensemble.predict` method of Code Block 12.9 does not modify the `weights` attribute, but instead defines a local variable that is normalized. This particular implementation is specifically for the `TimeForest` class, as normalizing the weights in place (i.e., as an attribute) would result in the incorrect ensemble when the next batch is trained.   ▷

The `TimeForest` class is implemented in Code Block 12.30. Here, we use `collections.deque`, which has the `append` and `popleft` methods. This is similar to the `queue.Queue` class used in Example 9.3, except we are explicitly tracking the `maxsize` using the `size` attribute of the parent `Ensemble` class.

```python
class TimeForest(Ensemble):

    def _getModel(self):
        self.models, self.weights = deque(), deque()
        self.gamma = self.params.get('gamma', 1)
        base_model = self.params.get('base_model', 'rf')
        if base_model == 'xgboost':
            return XGBoost
        else:
            return RandomForest # default

    def train(self, df, y):
        if self.type == 'classification' and not hasattr(self,
            'classes'):
            self.classes = np.sort(np.unique(y)).tolist()
        if not hasattr(self, 'vector_dim'):
            self.vector_dim = y.shape[1] if y.ndim == 2 else None
        model = self.Model(**self.model_params)
        model.train(df, y)
        if len(self.models) == self.size:
            self.models.popleft()
            self.weights.popleft()
        self.weights = deque([self.gamma * w for w in self.weights])
        self.models.append(model)
        self.weights.append(len(df))
```

Code Block 12.30: `TimeForest` implementation.

12.5.2 Survival Forests

Survival Problems

Next, we turn to a different temporal problem, that of training a model based on censored data. Recall from Definitions 5.6 and 5.7 the distinction between *censored* and *truncated* data: both involve data sets whose values are only known within a given interval; however, whereas the *counts* of censored instances are known in the former, they are completely unobserved in the latter. This distinction is clarified with two examples from user acquisition for an online product (e.g., app, website):

- *Censorship*—predicting user *churn*, which may be either an observed (e.g., for a subscription product, in which a user must actively cancel their membership) or unobserved (e.g., for a freemium product; see Seufert [2014]). This is an example of censorship, since we know the total user count and observe (or observe by a proxy metric, such as user hasn't logged in for three consecutive days). We also know the age of each user. For each user, then, we have a binary indicator expressing whether or not they have churned, and we have a time that represents lifetime for churned users and censorship time (e.g., current cohort age) for censored users.
- *Truncation*—predicting *payer rate* for a freemium product. This represents a truncation problem since, for any given cohort, we do not know *a priori* the number of users who will convert to payers, which is typically a small percentage of overall users. Users are also free to pay at any time, meaning that those who do eventually pay will have a distribution over the amount of time for first purchase. For a cohort that is seven days old, we therefore have a count of conversions, along with their pay-conversion timestamps, but we do not have a count of how many users have yet to convert.

Churn prediction is of course a core component of *customer lifetime-value* modeling (Section 10.3). In particular, the Bayesian mixture models we discussed therein had the serious deficit of not accounting for covariates. It is therefore of interest to see how tree methods might be retrofitted in the context of survival problems.

In fact, more generally, survival techniques are of interest any time behavior plays out over time; given the online nature of most web and mobile applications, in which users interact with virtual products over time, such problems proliferate industry.

Hazard and Survival

Recall that survival problems are primarily interested in determining or approximating a distribution for a random variable T that is temporal in nature. Typically, we may regard T as the *time to an event*; classically it is regarded as the *time to death* for patients in medical applications (i.e., literal *survival studies*), wherein the theory has its roots, or the *failure time* for engineering applications, or the *time to churn* or *customer lifetime* in customer lifetime-value models. The *survival function*

$$S(t) = \mathbb{P}(T > t) = 1 - F(t) \tag{12.21}$$

represents the probability that the individual is *still alive* at time t, and is a sort of dual to the cumulative distribution function F. Further, recall from that the *hazard function* (Definition 5.8) is defined by the ratio

$$h(t) = \frac{f(t)}{S(t)}, \tag{12.22}$$

which represents the differential probability of failing at time t, given that the device has yet to fail. Similarly, the *cumulative hazard function (CHF)* is defined as the integral

$$H(t) = \int_{-\infty}^{t} h(s)\,ds. \tag{12.23}$$

The CHF is a monotonically increasing function with range $[0, \infty)$. A simple calculation shows that

$$H(t) = -\ln(S(t)). \tag{12.24}$$

We leave the proof as an exercise for the reader.

Survival Trees

We next develop methodology that applies random forests to the survival setting to specifically solve problems involving right-censored data. There are several fruit methods[6] one might consider, such as reframing the survival problem as a classification problem or as a series of classification problems. The first true generalization of random forests to strictly adhere to the prescription of Breiman [2001], however, by taking into account the desired outcome in all aspects of formulating the forest, was Ishwaran, *et al.* [2008], who introduced the basic method of *random survival forests*.

Random Survival Forests

We follow Ishwaran, *et al.* [2008] in defining a random survival forest as follows.

Definition 12.13 (Random Survival Forest). *A* random survival forest *is a random forest (Definition 12.11) whose underlying decision trees use methods suitable for right-censored survival problems:*

---

[6] named after the proverbial *low-hanging fruit*.

1. *isHomogeneous—a data set is homogeneous if it consists of fewer than a fixed number of $d_0 > 0$ events;*
2. *bestSplit—returns the split that maximizes the* survival difference *between the child nodes;*
3. *getLabel—returns the Nelson-Aalen estimator for the cumulative hazard function.*

Note 12.5. The stopping criterion in Ishwaran, *et al.* [2008] is actually that each leaf node must have at least d_0 events, corresponding to a `min_samples_leaf` parameter. This is an appropriate definition, as it ensures there is sufficient data within each leaf to have a valid, though potentially weak, empirical estimator. We leave modification of the algorithm as an exercise for the reader.   ▷

We will discuss each of these method definitions in turn. Of particular interest with be the `bestSplit`, where we must decide how to determine the *survival difference* between two data sets.

Labels for Right-Censored Training Data

Before discussing the three method modifications, let us first align on notation for training labels, so that we might efficiently capture both observed lifetimes and censoring times for right-censored instances. Labels for our training data occur in pairs (t_i, δ_i), where t_i is a time and δ_i is the binary event indicator

$$\delta_i = \mathbb{I}[\text{event has occurred for instance } i].$$

Thus, if $\delta_i = 0$, we say that the ith instance is *censored* at time t_i, which we call the *censoring time*. In the context of marketing or product analytics, we say that the user is still *alive* or *active* at time t_i, which usually represents the *cohort time*, or *age*, of the user.

On the other hand, if $\delta_i = 1$, we say that we have observed the event for the ith instance, and we interpret the time t_i as the *event time* (or *survival time* or *customer lifetime*, etc.). In CLTV problems, the time t_i is the cohort time, or age, at which the user churned.

Stopping Criterion: Minimum Events in Leaf Nodes

The tree-growth stopping criterion is that the recursion should continue as long as a minimum number of events $d_0 > 0$ exists in each leaf. This will be a limiting factor on the `bestSplit` method (Algorithm 12.2), in that only qualifying splits should be considered. Furthermore, if no qualifying split can be found, we terminate the recursion in Algorithm 12.1 and return the current label.

Best-Split Criterion: Maximum Survival Difference

Four separate splitting criteria were discussed in Ishwaran and Kogalur [2007]: logrank splitting (based on the *Mantel-Cox test*, or *logrank test*), conservation-of-events splitting, logrank score splitting, and an approximate logrank splitting, which offers improved computational speed over the exact test.

We will focus on logrank splitting, which is derived from the *Mantel-Cox test*, which tests the null hypothesis that two groups have the same hazard function. Let $T_1 < \cdots < T_d$ represent the total pooled event times between the groups, Y_{ij} represent the number of individuals at risk in group i at time T_j, and E_{ij} represent the number of observed events in group i at time T_j. Further, let $Y_j = Y_{0j} + Y_{1j}$ and $E_j = E_{0j} + E_{1j}$. Then the Mantel-Cox test states that the *logrank statistic*

$$Z = \frac{\sum\limits_{j=1}^{d} \left(E_{ij} - \frac{Y_{ij} E_j}{Y_j} \right)}{\sqrt{\sum\limits_{j=1}^{d} \frac{Y_{ij} E_j}{Y_j} \left(\frac{Y_j - E_j}{Y_j} \right) \left(\frac{Y_j - Y_{ij}}{Y_j - 1} \right)}}, \tag{12.25}$$

is approximately distributed as a standard normal distribution for large samples, for both $i = 0, 1$. We leave the proof to the reader (see Exercise 12.13).

In the context of survival trees, the two groups represent the two resultant child nodes of a given split. The absolute logrank score $|Z|$ is a measure of survival difference between the two child nodes, with larger values representing better splits. The logrank splitting criterion therefore fixes $i = 0$ or 1 and then uses $-|Z|$ as an impurity metric for determining the best split.

To compute the logrank statistic, consider a set of survival data $\mathcal{D} = \{(x_i, t_i, \delta_i)\}_{i=1}^{n}$, which, without loss of generality, we may think of as the data belonging to a given branch node N_0. The random variables $T_1 < \cdots < T_d$ represent the ordered, distinct survival times $\{t_i | \delta_i = 1\}$. The quantities Y_j and E_j can be computed using

$$Y_j = \sum_{i=1}^{n} \mathbb{I}[t_i \geq T_j], \tag{12.26}$$

$$E_j = \sum_{i=1}^{n} \mathbb{I}[t_i = T_j] \delta_i, \tag{12.27}$$

respectively. Now, any decision rule $r : \mathbb{D} \to \mathbb{B}$ (Definition 12.1) will split the data into two subsets, based on the value of $r(x_i)$. We can therefore compute the quantities Y_{1j} and E_{1j} as

$$Y_{1j} = \sum_{i=1}^{n} r(x_i)\mathbb{I}[t_i \geq T_j], \tag{12.28}$$

$$E_{1j} = \sum_{i=1}^{n} r(x_i)\mathbb{I}[t_i = T_j]\delta_i, \tag{12.29}$$

respectively. Alternatively, to compute Y_{0j} and E_{0j}, we can simply replace $r(x_i)$ with $(1 - r(x_i))$ in the previous two equations.

Capturing Labels: Cumulative Hazard Function

Instead of numbers (regression) or class labels (classification), survival forests predict the survival function for a given instance, typically in the form of the cumulative hazard function CHF, which has better small-sample properties, making it ideal as a survival estimator within each leaf node.

If a given decision tree has a total of ℓ leaf nodes, we may view the tree τ as a mapping $\tau : \mathbb{D} \to \mathbb{Z}_\ell$ from feature vector into an ℓ-class encoding of the tree's leaves, defined via a tree-graph of individual decision rules. This further partitions the training data into ℓ subsets $\mathcal{D}_l = \{(x_i, t_i, \delta_i)\}_{i=1}^{n_l}$, for $l = 1, \ldots, \ell$, based on the condition that a datum (x_i, t_i, δ_i) falls into $\mathcal{D}_l$ if and only if $\tau(x_i) = l$.

Now, let $T_1 < \cdots < T_d$ be the pooled set of event times for the full training set, as we had before, and define

$$Y_{lj} = \sum_{i=1}^{n} \mathbb{I}[\tau(x_i) = l]\mathbb{I}[t_i \geq T_j], \tag{12.30}$$

$$E_{lj} = \sum_{i=1}^{n} \mathbb{I}[\tau(x_i) = l]\mathbb{I}[t_i = T_j]\delta_i, \tag{12.31}$$

which is equivalent to applying Equations (12.26) and (12.27) over the partial dataset $\mathcal{D}_l$, for $l = 1, \ldots, \ell$. The Nelson-Aalen estimator

$$\hat{H}_l(t) = \sum_{T_j \leq t} \frac{E_{lj}}{Y_{lj}} \tag{12.32}$$

thus yields an empirical estimate for the cumulative hazard function for the lth leaf, which further represents the prediction "label" for the given leaf.

Finally, for an arbitrary feature vector x, we have the survival tree's predicted cumulative hazard, given by

$$\hat{H}(t|x) = \hat{H}_{\tau(x)}(t); \tag{12.33}$$

i.e., the CHF associated with leaf $l = \tau(x)$.

Prediction: Ensemble Cumulative Hazard

Finally, suppose now that we have a forest comprised of m survival trees $\tau_1, \ldots, \tau_m$, with their hazard predictors $\hat{H}_1(t|x), \ldots, \hat{H}_m(t|x)$. The *ensemble* CHF for an arbitrary feature vector is therefore given by

$$\hat{H}(t|x) = \frac{1}{m} \sum_{j=1}^{m} \hat{H}_j(t|x). \tag{12.34}$$

Recall that each tree in a random forest is trained using a bootstrap sample of the original training data set $\mathcal{D}$. Thus, each survival tree, on average, excludes approximately 37% of the full training data. For the ith training datum (x_i, t_i, δ_i), let b_{ij} be the binary indicator representing whether (0) or not (1) the training instance was used to train the jth survival tree; i.e., $b_{ij} = 1$ if the ith datum was out-of-bag for the jth tree. The *out-of-bag* CHF for the ith training instance is therefore given by

$$\hat{H}_{oob}(t|x_i) = \frac{\sum\limits_{j=1}^{m} b_{ij} \hat{H}_j(t|x_i)}{\sum\limits_{j=1}^{m} b_{ij}}, \tag{12.35}$$

which can be used to measure out-of-bag accuracy of the ensemble.

Problems

12.1. How would one modify Algorithm 12.2 to allow for multiway splits for categorical features?

12.2. Implement the golden-search algorithm to determine the minimum value of the function $f(x) = -xe^{-x}$.

12.3. Update the _isHomogeneous method in Code Block 12.4 to provide a variance-based stopping criterion for regression problems.

12.4. Write a pruning method for the **DecisionTree** Python class.

12.5. Write an __add__(self, other) method for the **Ensemble** class (Code Block 12.9), following Definition 12.6, so that we can combine trained ensembles with the notation E + F.

12.6. Show that a bootstrap sample, on average, should contain $(1 - 1/e) \approx 63.21\%$ of the original data, when the sample is done with replacement and is the same size as the original sample. *Hint*: What is the probability of *not* selecting any particular datum from the set? Then take the limit $n \to \infty$ and recall the definition of e.

12.7. Compute the twelve predictions from the random forest shown in Figure 12.4 for the instance: sepal length 5.00, petal length 2.50, and petal width 2.00. What is the final vote tally for the three classes (setosa, versicolor, and virginica)?

12.8. Explain why the feature sex:M will never be used in the diabetes example of Code Block 12.17. *Hint*: consider carefully how the _bestSplit method of the DecisionTree class is written.

12.9. Explain the inherent problem in applying the AdaBoost Algorithm 12.6 to a classification problem with more than two classes. How might it be resolved?

12.10. Show that the starting model in the gradient boosting algorithm is the sample mean, when using squared error loss (i.e., prove Equation (12.16)).

12.11. Show that by summing the log loss Equation (12.15) over a set of data, one obtains the entropy impurity measure Equation (12.4).

12.12. Prove Equation (12.24).

12.13. Prove Equation (12.25). *Hint*: First, explain why $E_{ij} \sim \mathrm{HypGeom}(Y_j, E_j,)$ Then use known properties of the hypergeometric distribution and the central-limit theorem to prove the result.

Deep Thoughts

In this chapter, we discuss the basics of the *artificial neural network* (ANN) and *deep learning*, which constitute the foundation of modern artificial intelligence. Our favorite references are Aggarwal [2018] and Gèron [2019], the latter of which is focused on sci-kit learn and tensorflow. For a short introduction of neural networks, see Efron and Hastie [2016]. Another classic text on the subject is Goodfellow, *et al.* [2016]. For additional references on deep learning in Python, see Chollet [2021] or Stevens, *et al.* [2020].

The idea of an artificial neuron was first introduced in the context of propositional logic by McColloch and Pitts [1943]; here, the proposed artificial neuron activated its output only if a sufficient number of binary inputs were activated. The authors showed that is possible to build a network of such neurons that could compute arbitrary logical propositions in a systematic way.

Hebb's rule provided another crucial clue to the development of the theory: whenever one biological neuron triggers another, the connection between the two grows stronger; i.e., *cells that fire together wire together* (see Hebb [1949]). This observation was a key inspiration to the *perceptron learning algorithm*, introduced a few years later.

The *perceptron* (also known as a *linear threshold unit* (LTU)), introduced by Rosenblatt [1957], is a variation of the purely logical neurons introduced years earlier and constitutes the basis of modern ANNs. Reading from this paper:

> Recent theoretical studies by this writer indicate that it should be feasible to construct an electronic or electromechanical system which will learn to recognize similarities or identities between patterns of optical, electrical, or tonal information, in a manner which may be closely analogous to the perceptual process of a biological brain. The proposed system depends on probabilistic rather than deterministic principles for its operation, and gains its reliability from the properties of statistical measurements obtained from

large populations of elements. A system which operates according
to these principles will be called a *perceptron*.

Yes, that *should* be feasible enough. Let's ask Siri...

13.1 Artificial Neurons

We begin with a brief discussion of the simplest neural architecture: the
perceptron. We then introduce the perceptron learning algorithm, how perceptrons can be generalized to multiclass classification problems, and conclude with the generalization of the perceptron to general artificial neurons.
This will form the basis for our discussion on neural networks in our next
section.

13.1.1 The Perceptron

The *perceptron* is the simplest neural network, which consists of a number
of inputs (the *input layer*) and a single output node.

Definition 13.1. *A* perceptron, *or linear threshold unit (LTU), is a binary
classifier over a set of input features* $x \in \mathbb{R}^d$ *that is defined by the composite
function* $f = H \circ \omega$, *where* $\omega : \mathbb{R}^d \to \mathbb{R}$ *is the linear mapping*

$$\omega(x) = \beta + \sum_{j=1}^{d} w_j x_j, \tag{13.1}$$

where the parameter β *is referred to as the* bias, *the values* $\{w_j\}_{j=1}^{d}$ *are
given weights, and* $H(x) = \mathbb{I}[x \geq 0]$ *is the classic Heaviside function.*

At face value, this seems right in line with the generalized linear models
we discussed in Chapter 7. For instance, if we replaced H with the logistic function, we would have the functional model for logistic regression; if
we replaced it with an identity, we'd have the functional form for linear

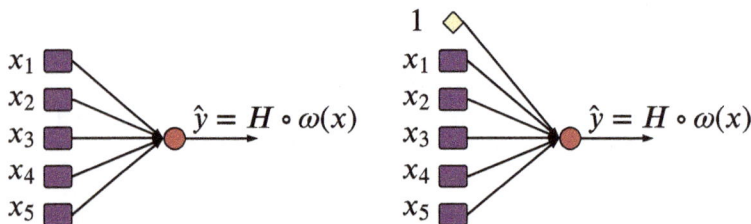

Fig. 13.1: The perceptron with (right) and without (left) a bias.

regression. The rational for considering the Heaviside function is based in neuroscience: neurons in the brain are connected to other cells via synapses, which may have a particular weight; the given neuron, then, fires an electrical signal via its axon only if the weighted input signal passes a particular threshold. This analogy can be better understood by representing the perceptron using a graphical model, as shown in Figure 13.1.

The perceptron is represented by the red circle, whereas the inputs (from other cells) are represented by the purple squares. The diamond node (Figure 13.1 (right)) represents the optional bias term. (We may, without loss of generality, represent the bias term as an additional *trivial neuron* with constant value 1 and weight β.)

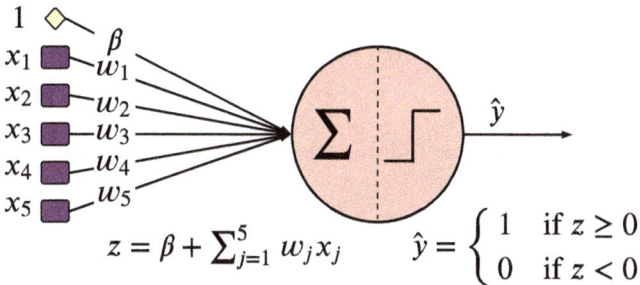

$$z = \beta + \sum_{j=1}^{5} w_j x_j \qquad \hat{y} = \begin{cases} 1 & \text{if } z \geq 0 \\ 0 & \text{if } z < 0 \end{cases}$$

Fig. 13.2: An exploded schematic of a perceptron with bias.

In order to understand the inner workings of the perceptron, it is helpful to consider the exploded view shown in Figure 13.2. Here, we see that the perceptron follows a two-step process in determining its output. First, it computes the inner product $z = w(x)$ of its inputs and weights. Second, it determines if the cumulative weight z is sufficient to warrant an output signal; i.e., $H(z) = 1$ when the cumulative weighted inputs are sufficient for an output to fire, whereas $H(z) = 0$ otherwise.

As shown in Figure 13.2, it is often useful to think of the model bias as an additional input x_0, which is set to $x_0 = 1$. In this case, we can set $w_0 = \beta$, such that

$$w(x) = \beta + \sum_{j=1}^{d} w_j x_j = \sum_{j=0}^{d} w_j x_j.$$

Given a set of weights, the level sets of the function $w(x)$ define hyperplanes in $\mathbb{R}^d$, meaning that the perceptron has a linear decision boundary, similar to logistic regression. (For data that are not linearly separable, we require more advanced neural networks.)

13.1.2 The Perceptron Learning Algorithm

In order to train a perceptron model, we need to determine the weights $w_0, \dots, w_d$ that minimize the binary loss function (Equation (6.19)), which is equivalent to the squared-error loss (Equation (6.17)), since the target variable is binary. Given a training data set $\mathcal{D} = \{(x_i, y_i)\}_{i=1}^n$, the cumulative binary loss is therefore equivalent to

$$\sum_{\mathcal{D}} L(y, \hat{y}) = \sum_{\mathcal{D}} (y - \hat{y})^2 = \sum_{\mathcal{D}} (y - H \circ \omega(x))^2. \tag{13.2}$$

Since this represents a staircase function, its derivative vanishes, except for a countable number of points at which it diverges to infinity. Nevertheless, we can construct an algorithm that mimics stochastic gradient descent, as given by Algorithm 13.1. The algorithm typically continues until the perceptron has achieved a minimum accuracy. Convergence is guaranteed if the training instances are linearly separable (Rosenblatt [1957]), otherwise, the algorithm will never achieve perfect accuracy.

In order to motivate the update rule of line 6, let us formally differentiate the loss function as follows

$$\frac{\partial L}{\partial w_j} = \frac{\partial L}{\partial \hat{y}} \frac{\partial \hat{y}}{\partial z} \frac{\partial z}{\partial w_j} = -2(y - \hat{y}) \delta(z) x_j$$

where $\delta(z)$ is the Dirac delta function (see Section 1.3), which, the reader will undoubtedly recall, may be viewed as the derivative of the Heaviside function. The delta function is problematic, as it either diverges (whenever the input vector is orthogonal to the weights, so that $\sum_{j=0}^d w_j x_j = 0$) or vanishes otherwise. Nevertheless, the perceptron learning algorithm functions as expected if we remove the delta function in the gradient and absorb the factor of 2 into the learning rate η.

Algorithm 13.1: The perceptron learning algorithm.

Input: labeled data $\mathcal{D} = \{(x_i, y_i)\}_{i=1}^n$; $x_i \in \mathbb{R}^d$ and $y_i \in \mathbb{B}$;
      learning rate η
Output: The set $\{w_j\}_{j=0}^d$ of perceptron weights, with bias $w_0 = \beta$.

1 Set $w_j = 0$ for $j = 0, \dots, d$
2 Set *converged* = **False**
3 **while not** *converged* **do**
4   Randomly select $(x, y) \in \mathcal{D}$
5   Compute $\hat{y} = H \circ \omega(x)$
6   Update $w_j = w_j + \eta(y - \hat{y})x_j$, for $j = 0, \dots, d$
7   *converged* = checkConverged$(\mathcal{D}, \{w_j\})$
8 **end**
9 **return** $w_0, w_1, \dots, w_d$

13.1.3 Vector-output Perceptrons

If the target variable is a c-dimensional binary vector ($y \in \mathbb{B}^c$), we can generalize our model by stacking c individual linear threshold units into a *layer*, which may be regarded as a *vector perceptron*. In this case, each individual LTU has its own set of weights. Such an arrangement is shown in Figure 13.3.

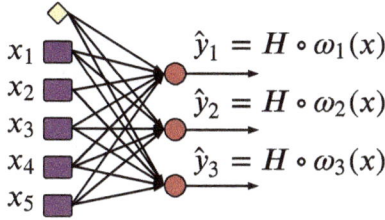

$$x_1, x_2, x_3, x_4, x_5$$

$$\hat{y}_1 = H \circ \omega_1(x)$$
$$\hat{y}_2 = H \circ \omega_2(x)$$
$$\hat{y}_3 = H \circ \omega_3(x)$$

Fig. 13.3: A vector perceptron for target variable $y \in \mathbb{B}^c$, shown for $c = 3$.

For the ith perceptron in a perceptron vector, we may regard the function ω_i as the *one-form*[1]

$$\omega_i(x) = \sum_{j=0}^{d} w_{ij} x_j.$$

We may further stack these one-forms as rows in a matrix W, with components

$$W = \begin{bmatrix} w_{11} & \cdots & w_{1d} \\ \vdots & \ddots & \vdots \\ w_{c1} & \cdots & w_{cd} \end{bmatrix}, \tag{13.3}$$

where w_{ij} is the weight between the jth input and the ith output, for $i = 1, \ldots, c$ and $j = 1, \ldots d$. Note that many authors define the weight matrix as the *transpose* of this definition; we, however, feel it to be more natural to define our weight matrix so that $z = Wx$, as we move forward through the neural net.

Similarly, we may regard $\beta \in \mathbb{R}^c$ as a vector, so that the operations of a given vector perceptron may be captured in a single vector equation

$$z = \beta + Wx$$
$$\hat{y} = H(z),$$

where $x \in \mathbb{R}^d$ and $H : \mathbb{R}^c \to \mathbb{R}^c$ operates on its vector input componentwise.

---

[1] A one-form can intuitively thought of as a mapping from a vector space into the real numbers; formally, this may be regarded as a *row vector*, since the matrix product of a row vector and a (column) vector is a scalar.

13.1.4 Softmax Neurons

Obviously, our definition of the vector perceptron is not readily applicable to multiclass classification problems, as the output of the former is expressed as an arbitrary binary vector. To remedy this, we will apply a *softmax layer*, which will normalize the pre-outputs into a set of probabilities. This will require two special definitions.

Definition 13.2. *A Σ neuron (sigma neuron) is a perceptron for which the Heaviside function is replaced with the identity mapping $\iota(z) = z$.*

A level i softmax neuron, for $i = 1, \ldots, c$, where c is the number of inputs, is the ith component of the softmax transform (Equation (6.12)),

$$s_i(z) = \frac{e^{z_i}}{\sum_{j=1}^{c} e^{z_j}}, \tag{13.4}$$

where $z \in \mathbb{R}^c$ is the input vector.

Graphically, we will represent sigma neurons as circles with embedded Σ signs and softmax neurons as hexagons. A sigma neuron may be viewed as the first half of the perceptron.

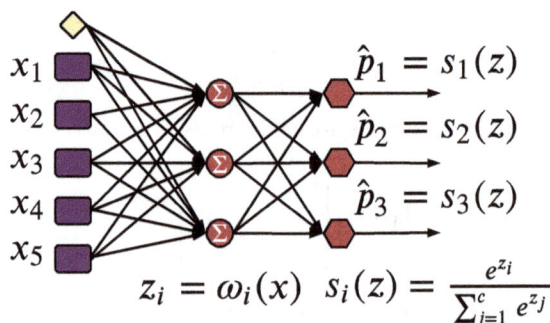

$$z_i = \omega_i(x) \quad s_i(z) = \frac{e^{z_i}}{\sum_{j=1}^{c} e^{z_j}}$$

Fig. 13.4: Softmax transform.

Sigma and softmax neurons always occur in pairs, as shown in Figure 13.4. The reason for separating these, as opposed to smashing them together into a single neuron, is that each softmax neuron requires the output from each sigma neuron. The final outputs $\hat{p}_1, \ldots, \hat{p}_c$ may then be viewed as probabilities for the c-class classification problem.

Though separated from a diagrammatic perspective, in the end we will program them concurrently in a single neural layer called a *softmax layer*. Figure 13.4 may therefore be regarded as simply an exploded view into such a layer.

13.1.5 Activation Functions

For general artificial neural networks, we will generalize the perceptron into a closely related concept called the *artificial neuron*. Artificial neurons still consist of two steps: a summation step and an activation step. However, the activation step may be achieved with a broader class of functions. In particular, we will replace the Heaviside function H with a general *activation function* ϕ.

We previously saw that the lack of differentiability of the Heaviside function created problems when attempting to calculate the gradient of the loss function. For general ANNs, we will be interested in deploying stochastic gradient descent to learn the network weights. As such, we will require both the functional form and the derivative of any activation function we use. In this section, we will highlight several commonly used forms for the activation function. Plots of these functions are shown in Figure 13.5; their derivatives are plotted in Figure 13.6.

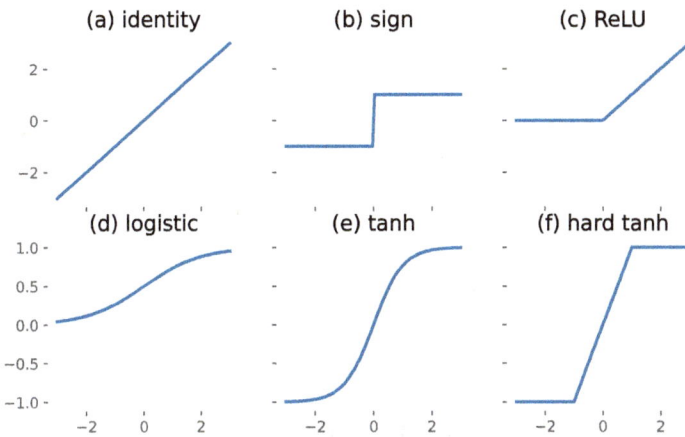

Fig. 13.5: Commonly used activation functions.

Identity Function

The identify function and its derivative are given by

$$\phi(z) = z$$
$$\phi'(z) = 1.$$

An artificial neuron with the identity activation is simply a *sigma neuron* (Definition 13.2). The chief usage here is in coupling the neuron with a softmax output layer.

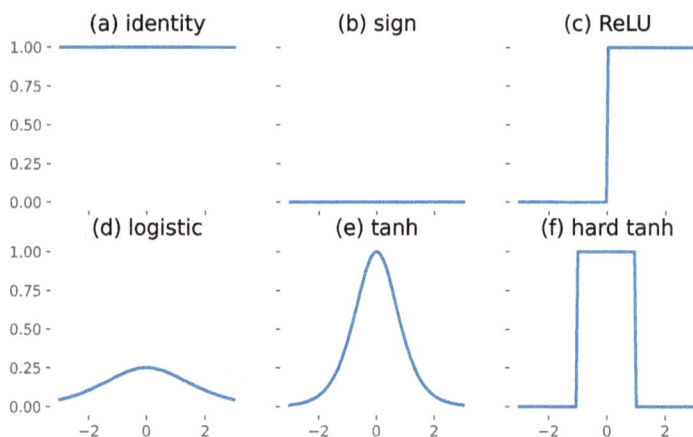

Fig. 13.6: Derivatives of commonly used activation functions.

Sign Function

The sign function and its derivative are given by

$$\phi(z) = \text{sign}(z) = H(z) - H(-z)$$
$$\phi'(z) = \text{sign}'(z) = \delta(z).$$

The derivative of this function vanishes everywhere, except at the origin, where it diverges. It is therefore rarely used in practice.

Rectified Linear Unit

.

The *rectified linear unit (ReLU)* and its derivative are given by

$$\phi(z) = \max\{0, z\} = zH(z)$$
$$\phi'(z) = H(z)$$

Despite the lack of differentiability at $z = 0$ and the vanishing of the derivative for $z < 0$, the ReLU function tends to work well in practice, and is perhaps the most commonly used form of activation. It is also relatively easy to compute. Given the nonnegative aspect of the ReLU function, it is also commonly used as the activation function in the output layer for the regression of nonnegative target variables.

Logistic Function

The *logistic function* and its derivative are given by

$$\phi(z) = \frac{1}{1 + \exp(-z)}$$

$$\phi'(z) = \frac{\exp(-z)}{(1 + \exp(-z))^2} = \phi(z)\,[1 - \phi(z)]$$

The second equation here is the *logistic equation*, and is a result of the functional form of the derivative. This is one of the classic activation functions, but has largely been replaced by the ReLU function in modern practice.

Hyperbolic Tangent Function

The and its derivative are given by

$$\phi(z) = \tanh(z) = \frac{\exp(2z) - 1}{\exp(2z) + 1}$$

$$\phi'(z) = \tanh'(z) = \frac{4\exp(2z)}{(1 + \exp(2z))^2} = \mathrm{sech}^2(z) = 1 - \tanh(z)^2$$

Though the hyperbolic tangent is a simple transformation of the logistic function, the hyperbolic tangent is preferred when the final outputs might be both positive or negative. Additionally, the fact that it is centered and has a larger gradient tends to make it easier to train in practice.

"Hard" Hyperbolic Tangent Function

The *hard hyperbolic tangent function* and its derivative are given by

$$\phi(z) = \max\{-1, \min\{1, z\}\}$$
$$\phi'(z) = \mathbb{I}[|z| < 1] = H(1 - z) - H(-1 - z)$$

Similar to how ReLU has largely replaced the logistic function, the hard hyperbolic tangent has largely replaced the regular (soft) hyperbolic tangent in modern usage. This activation has similar scaling to its soft counterpart, but has the advantage of a piecewise-constant derivative.

Nonlinear Activations

Finally, we note the importance of nonlinear activation functions, without which the point of constructing neural network architectures would be lost. This is because of the LC^3 theorem: linear combinations of linear combinations of linear combinations are linear combinations. In other words, if we were to build a network that only consisted of layer after layer of linear combinations, we would, in the end, only have an over specified linear combination.

13.2 Neural Networks

In this section, we generalize the perceptron to more complex graphical patterns known as artificial neural networks. We begin by laying out the basic definitions and nomenclature. Then we discuss training neural networks using stochastic gradient descent to minimize a given loss function.

13.2.1 Neural Networks

To begin, we formalize our concept of a *neural layer*.

Definition 13.3. *A* neural layer *(or, simply,* layer*) is a set of artificial neurons that share a set of inputs. The* size *of a neural layer is the number of* nontrivial neurons *(i.e., non-bias neurons) that comprise it.*

For example, the red circular output neurons shown in Figure 13.3 constitute a layer. Similarly, the blue rectangles constitute a special layer known as an *input layer*.

Definition 13.4. *An* artificial neural network (ANN) *is a sequence of neural layers, such that the inputs to one layer are the outputs of the preceding layer.*

In an artificial neural network, the input layer *consists of the set of neurons that represent the direct functional input of the network, the* output layer *consists of the set of neurons that represent the final functional output of the network, and the* hidden layers *is the set of layers that are sandwiched between.*

An artificial neural network is said to be shallow *if it only consists of zero, one, or two hidden layers. Otherwise it is said to be* deep.

For example, in Figure 13.4, the hexagonal neurons constitute the output layer. In this case, the output layer is a special kind of output known as a *softmax layer*. *Ibid.*, the middel sigma neurons constitute a hidden layer.

An example of a deep neural network consisting of four layers is shown in Figure 13.7. (The input layer is typically not counted.) Each layer and its input connections are colored together. Here, we have a two-node output layer ($l = 4$), a five-node input layer ($l = 0$), and three hidden layers ($l = 1, 2, 3$) of sizes five, four, and three, respectively.

Neural Network Notation Convention

To describe a general artificial neural network, we will make use of the following notation convention.

Consider a neural network with m layers (not counting the input layer), indexed by l. The size of layer l will be denoted by r_l. We therefore have:

1. layer $l = 0$ is the input layer, with size $r_0 = d$;

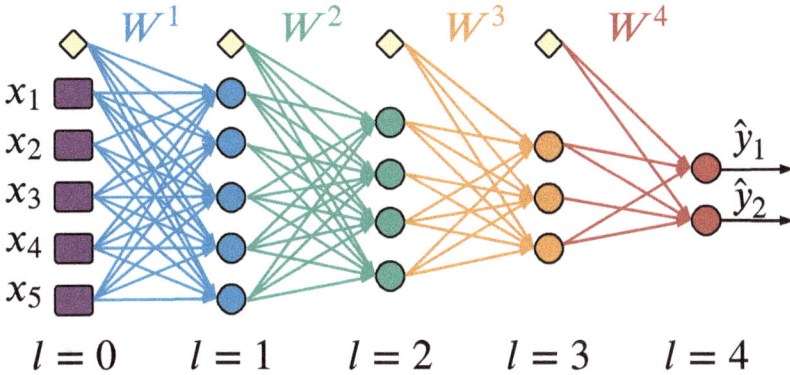

Fig. 13.7: An artificial neural network.

2. layers $l = 1, \ldots, m-1$ constitute the hidden layers, with sizes $r_1, \ldots, r_{m-1}$;
3. layer $l = m$ is the output layer, with size $r_m = c$.

We will assume, as often is the case, that each layer uses a common activation function within the layer. We will denote the activation function for layer l as $\phi_l : \mathbb{R} \to \mathbb{R}$.

Aside from the activation function, the variables of layer l will be marked with a *superscript* l^2.

As was the case with vector perceptrons, the weights for all connections leading into layer l may be represented by a single $r_l \times r_{l-1}$ matrix W^l and a vector $\beta^l \in \mathbb{R}^{r_l}$, for $l = 1, \ldots, m$, which takes the form of Equation (13.3) (with d replaced by r_{l-1} and c replaced by r_l):

$$W^l = \begin{bmatrix} w_{11}^l & \cdots & w_{1r_{l-1}}^l \\ \vdots & \ddots & \vdots \\ w_{r_l 1}^l & \cdots & w_{r_l r_{l-1}}^l \end{bmatrix}. \tag{13.5}$$

We further added superscripts to the weights and biases to indicate their respective layer number. In particular:

1. the quantity β_i^l represents the bias of the ith neuron of layer l;
2. the quantity w_{ij}^l represents the weight between the ith neuron of layer l and the jth neuron of layer $l - 1$, for $i = 1, \ldots, r_l$ and $j = 1, \ldots, r_{l-1}$.

Finally, we denote the output of the ith neuron of layer l with the variable a_i^l, choosing the variable a as these values correspond to the "activated" values. Naturally, we identify $a_i^0 = x_i$ for the input layer. For layers

---

[2] We use a subscript, as opposed to a superscript, for the activation function for two reasons: first, no further enumeration is needed and, second, so that we may more cleanly represent the derivative $\phi_l'(z)$.

$l = 1, \ldots, m$, we have the intermediary result $z^l = \beta^l + W^l a^{l-1}$, such that the ith component of this vector is given by

$$z_i^l = \beta_i^l + \sum_{j=1}^{r_{l-1}} w_{ij} a_j^{l-1}.$$

The output of the neuron is then given by the activated value $a_i^l = \phi_l(z_i^l)$, as usual. When we define $\phi_l : \mathbb{R}^{r_l} \to \mathbb{R}^{r_l}$ as a vector function, we simply allow the activation function to act elementwise for components $i = 1, \ldots, r_l$.

Feedforward Computation

Though the notation itself reveals the precise computational steps required to compute the output of a neural network with a given set of weights and a given input vector, we define the following *feedforward computation algorithm* in Algorithms 13.2 and 13.3 for clarity. In the former, we represent the equations in their vector–matrix form, whereas in the latter, we explicitly represent the calculations for each individual neuron in the network.

13.2.2 Backpropagation

In order to determine the weights for an artificial neural network that minimize a given loss function over a set of training data, we will apply stochastic or mini-batch gradient descent, as usual. The challenge for us, however, is in determining the gradients in the first place. The key insight is that neural networks are, essentially, intrinsic composite functions that are defined by a computational graph. As such, we may rely on rules of calculus to arrive at expressions for the various derivatives.

Algorithm 13.2: The feedforward algorithm (matrix version).

Input: an input vector $x \in \mathbb{R}^d$;
        a given neural network of size m;
        a set of m activation functions $\phi_1, \ldots, \phi_m$;
        a given set of (current) bias vectors $\mathcal{B} = \{\beta^1, \ldots, \beta^m\}$;
        a given set of (current) weights $\mathcal{W} = \{W^1, \ldots, W^m\}$
Output: Vector of predicted values $\hat{y} \in \mathbb{R}^c$.
1   Set vector $a^0 = \langle x_1, \ldots, x_d \rangle$
2   **for** *each layer* $l = 1, \ldots, m$ **do**
3   Compute $z^l = \beta^l + W^l a^{l-1}$
4   Compute $a^l = \phi_l(z^l)$
5   **end**
6   **return** $\hat{y} = a^m \in \mathbb{R}^c$

Gradients *à la* Chain Rule

The loss function may be regarded as a final output value that comes after the (potentially) multiclass output layer. As such, computing its gradient reduces to the ability to compute arbitrary partial derivatives in directed acyclic computational graphs. (Recall Definition 8.4.) By a *computational graph*, we mean a graph whose nodes represent variables, and whose directed edges represent functional relations, similar to our definition of a causal structure (Definition 8.7). As it turns out, the chain rule takes a simple form for such a use case, though it is exponentially expensive to compute as the size of the graph grows. Our present goal is merely to introduce the full chain rule and demonstrate its prohibitive cost. In the following section, we introduce a technique known as backpropagation that is used in practice.

The Chain Rule for Computational DAGs

Before proceeding, recall that a path P in a graph $\mathcal{G}$ between two connected points x and y is a sequence of points $\{v_0^P, v_1^P, \ldots, v_{\ell(P)}^P\} \subset \mathcal{G}$, where $\ell(P)$ is the length of the path, such that $v_0^P = x$, $v_{\ell(P)}^P = y$, and

$$v_{i-1}^P \in \mathrm{par}(v_i^P),$$

for $i = 1, \ldots, \ell(P)$. Given this definition, we may write the following. (Though it is stated in terms of general computational DAGs, one should be content to picture the graph as a neural network for our present purpose.)

Algorithm 13.3: The feedforward algorithm (scalar version).

Input: an input vector $x \in \mathbb{R}^d$;
  a given neural network of size m;
  a set of m activation functions $\phi_1, \ldots, \phi_m$;
  a given set of (current) bias vectors $\mathcal{B} = \{\beta^1, \ldots, \beta^m\}$;
  a given set of (current) weights $\mathcal{W} = \{W^1, \ldots, W^m\}$
Output: The predicted values $\{\hat{y}_i\}_{i=1}^c$.

1 Set vector $a_i^0 = x_i$ for $i = 1, \ldots, d$
2 **for** *each layer* $l = 1, \ldots, m$ **do**
3 $\quad$ **for** *each neuron* $i = 1, \ldots, r_l$ **do**
4 $\qquad$ Compute $z_i^l = \beta_i^l + \sum\limits_{j=1}^{r_{l-1}} w_{ij}^l a_j^{l-1}$
5 $\qquad$ Compute $a_i^l = \phi_l(z_i^l)$
6 $\quad$ **end**
7 **end**
8 Set $\hat{y}_i = a_i^m$, for $i = 1, \ldots, c$
9 **return** $\hat{y}_1, \ldots, \hat{y}_c$

Theorem 13.1 (Chain Rule for Computational Graphs). *Let $\mathcal{G}$ be a directed acyclic computational graph, the nodes $x, y \in \mathcal{G}$ with $x \in \mathrm{anc}(y)$, and $\mathcal{P}$ the set of all directed paths that connect x to y. Then the partial derivative of y with respect to x may be expressed as*

$$\frac{\partial y}{\partial x} = \sum_{P \in \mathcal{P}} \left(\prod_{i=1}^{\ell(P)} \frac{\partial v_i^P}{\partial v_{i-1}^P} \right). \tag{13.6}$$

Classic Chain Rule Examples with their Corresponding DAGs

We next consider the classic chain rule from univariate and multivariate calculus, and show their equivalency to Theorem 13.1.

Example 13.1. Consider the chain (Definition 8.12) shown in Figure 13.8. This represents the graphical representation of the classic chain rule from

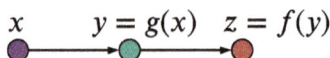

Fig. 13.8: A simple chain, representing $z = f(g(x))$.

calculus, in which we have a composite function $z = f(y)$ and $y = g(x)$; or, in short, $z = f(g(x))$. From basic calculus, we have

$$\frac{dz}{dx} = f'(g(x))g'(x) = \frac{dz}{dy}\frac{dy}{dx},$$

which is the simplest application of Equation (13.6), as there is only a single directed path that connects x to z.   ▷

Example 13.2. As a simple example of a multivariate chain rule, consider the computational DAG shown in Figure 13.9. Here, we consider the mul-

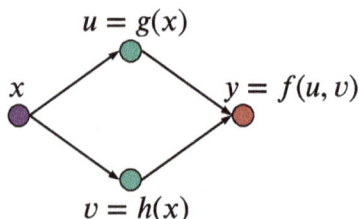

Fig. 13.9: A fork coupled with a collider, representing $y = f(g(x), h(x))$.

tivariate composite function $y = f(u, v)$, $u = g(x)$, and $v = h(x)$. In the graphical representation for this composite function, there are two paths linking x to y. From calculus, we have

$$\frac{dy}{dx} = f_u(g(x), v(x))g'(x) + f_v(g(x), v(x))v'(x) = \frac{\partial y}{\partial u}\frac{du}{dx} + \frac{\partial y}{\partial v}\frac{dv}{dx}.$$

One can readily verify this is equivalent to the formula produced by Equation (13.6).   ▷

Application of Chain Rule to ANNs

The graphical representation of a neural network is not a fully expressed computational graph, but a shorthand, since each neuron is actually a representation of a short chain and since the weights are not explicitly represented as nodes (though they are variables). To see this, consider the shallow ANN of Figure 13.10 comprised of a single input, no bias, a three-node hidden layer, and a single node output. We can easily explode this into a full computational graph, including the final loss function (which is a function of both the final output $\hat{y}$ and the true label y of the training instance), as shown in Figure 13.11.

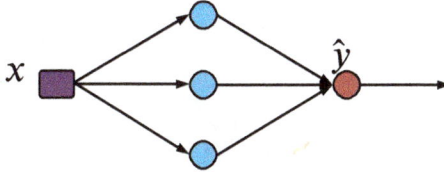

Fig. 13.10: A shallow neural network with two layers.

Naturally, if we had more than one input, our problems would multiply, as we would require a unique weight node for each *connection*: for example, five inputs into a six-node layer would require thirty weight nodes (thirty-six if we include a bias). Nevertheless, we are performing this exercise of the exploded computational graph simple to gain some insight into the workings of more general neural networks.

Each of the six weights shown in Figure 13.11 has only a single directed path to the loss function L. In order to examine a multipath application of the chain rule, we must, minimally, add an additional neuron to the output layer. The exploded computational graph corresponding to this change is shown in Figure 13.12.

By means of example, let us construct the partial derivative of the loss function with respect to the first weight w_{11}^1. There are two paths between

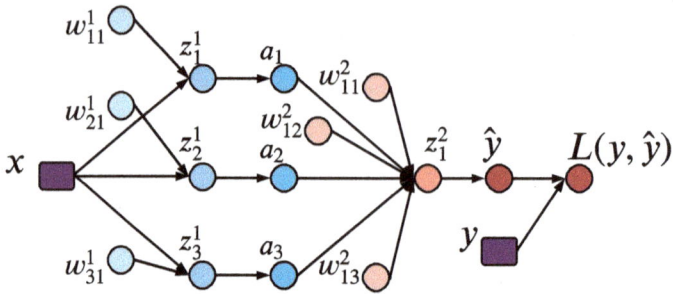

Fig. 13.11: The full computational DAG corresponding to the ANN of Figure 13.10.

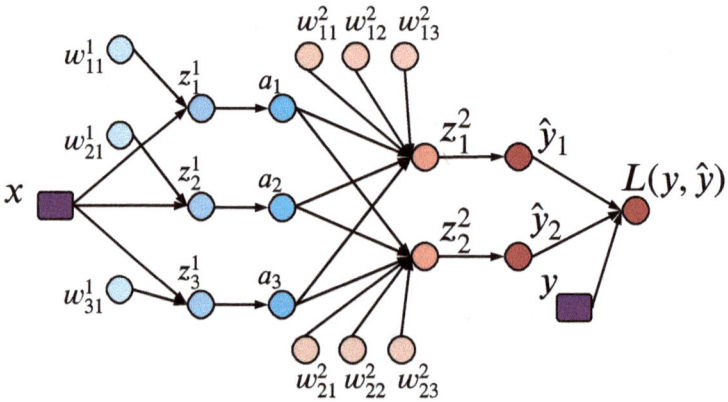

Fig. 13.12: The full computational DAG corresponding of a two-layer ANN with two output nodes.

these two endpoint nodes, one passing through each output neuron. Let us start with the final segment of the first path:

$$\frac{\partial L}{\partial z_1^2} = \frac{\partial L}{\partial \hat{y}_1} \frac{d\hat{y}_1}{dz_1^2} = \frac{\partial L}{\partial \hat{y}_1} \phi_2'(z_1^2),$$

where ϕ_2 is the activation function of the second (output) layer.

Next, let us consider the connection between a_1 and z_1^2. Since the quantity z_1^2 is simply the inner product

$$z_1^2 = w_{11}^2 a_1 + w_{12}^2 a_2 + w_{13}^2 a_3$$

between the inputs (a_1, a_2, a_3) and the weights $(w_{11}^2, w_{12}^2, w_{13}^2)$, we find that

$$\frac{\partial z_1^2}{\partial a_1} = w_{11}^2.$$

Finally, we have

$$\frac{\partial a_1}{\partial w_{11}^1} = \frac{da_1}{dz_1^1} \frac{\partial z_1^1}{\partial w_{11}^1} = \phi'(z_1^1)x,$$

since $z_1^1 = w_{11}^1 x$.

Combining the preceding results and repeating the calculation for the second path, we have

$$\frac{\partial L}{\partial w_{11}^1} = \frac{\partial L}{\partial \hat{y}_1} \frac{d\hat{y}_1}{dz_1^2} \frac{\partial z_1^2}{\partial a_1} \frac{da_1}{dz_1^1} \frac{\partial z_1^1}{\partial w_{11}^1} + \frac{\partial L}{\partial \hat{y}_2} \frac{d\hat{y}_2}{dz_2^2} \frac{\partial z_2^2}{\partial a_1} \frac{da_1}{dz_1^1} \frac{\partial z_1^1}{\partial w_{11}^1}$$

$$= \frac{\partial L}{\partial \hat{y}_1} \phi_2'(z_1^2) w_{11}^2 \phi_1'(z_1^1)x + \frac{\partial L}{\partial \hat{y}_2} \phi_2'(z_2^2) w_{21}^2 \phi_1'(z_1^1)x.$$

Voila! Couldn't be easier!

Exponential Complexity

The problem with the chain rule, as expressed by Equation (13.6), is that the number of paths increases *exponentially* with the depth of the graph. For a graph with m layers, of sizes $r_1, \ldots, r_m$, respectively, the number of paths between the loss function and any given first-layer weight is the product

$$\text{number of paths} = \prod_{l=2}^{m} r_l.$$

Since there are r_1 neurons in the first layer, we require a total of $\prod_{l=1}^{m} r_l$ path computations in total. The total number of paths we must consider to compute the full gradient is therefore given by

$$\text{total number of paths} = \sum_{l'=1}^{m} \prod_{l=l'}^{m} r_l.$$

However, as should be clear from our discussion of Figure 13.12, there is much redundancy in these calculations. As it turns out, we can construct an algorithm using *dynamic programming* that renders the total computational complexity as merely linear with depth, not exponential. We take up this topic forthwith.

The Backpropagation Algorithm

As we have seen, the number of paths between any given weight and the loss function increases exponentially with depth, making the brute-force application of the chain rule prohibitively expensive. In this section, we show how *dynamic programming* can be used to compute the full gradient of the loss function with respect to the full set of weights. The resulting algorithm is known as *backpropagation* and is given in Algorithm 13.4.

Algorithm 13.4 works because it exploits the redundancy in the calculation of the individual partial derivative. Moreover, the number of computations scales like the number of neurons, not exponentially like the number of layers. This is because we start at the end and work backwards, saving the intermediary partial derivatives as we go.

Notice that the variables δ_i^l are simply the partial derivatives of L with respect to the sigma variables z_i^l. The reason for focusing on these values, in particular, is that they are used in two ways. First, to compute the partial derivative of L with respect to any of the r_{l-1} weights associated with the ith neuron of the lth layer, we take

---

Algorithm 13.4: Backpropagation (scalar version).

Input: a training instance (x, y);
  a given neural network of size m;
  a set of m activation functions $\phi_1, \ldots, \phi_m$;
  a given set of (current) bias vectors $\mathcal{B} = \{\beta^1, \ldots, \beta^m\}$;
  a given set of (current) weights $\mathcal{W} = \{W^1, \ldots, W^m\}$

Output: The full gradient $\nabla_{\mathcal{W}} L$ and $\nabla_{\mathcal{B}} L$.

1 Compute $\hat{y} = f(x, \mathcal{W})$ using the feedforward algorithm, saving the values of all intermediary variables

2 **for** *each layer* $l = m, \ldots, 1$ *(in descending order)* **do**

3   **for** *each neuron* $i = 1, \ldots, r_m$ **do**

4   **if** *layer* l *is output layer* $l = m$ **then**

5   Set $\delta_i^m = \dfrac{\partial L}{\partial z_i^m} = \dfrac{\partial L}{\partial \hat{y}_i} \phi_m'(z_i^m)$

6   **else**

7   Set $\delta_i^l = \dfrac{\partial L}{\partial z_i^l} = \left(\displaystyle\sum_{k=1}^{r_{l+1}} w_{ki}^{l+1} \delta_k^{l+1} \right) \phi_l'(z_i^l)$

8   **end**

9   Set $\dfrac{\partial L}{\partial w_{ij}^l} = \delta_i^l a_j^{l-1}$ for $j = 1, \ldots, r_{l-1}$

10   Set $\dfrac{\partial L}{\partial \beta_i^l} = \delta_i^l$

11   **end**

12 **end**

13 **return** $\nabla_{\mathcal{W}} L = \left\{ \dfrac{\partial L}{\partial w_{ij}^l} \right\}_{l=1 \ i=1 \ j=1}^{m \quad r_l \quad r_{l-1}}$ and $\nabla_{\mathcal{B}} L = \left\{ \dfrac{\partial L}{\partial \beta_i^l} \right\}_{l=1 \ i=1}^{m \quad r_l}$

$$\frac{\partial L}{\partial w_{ij}^l} = \frac{\partial L}{\partial z_i^l} \frac{\partial z_i^l}{\partial w_{ij}^l} = \delta_i^l a_j^{l-1}$$

$$\frac{\partial L}{\partial \beta_i^l} = \frac{\partial L}{\partial z_i^l} \frac{\partial z_i^l}{\partial \beta_i^l} = \delta_i^l,$$

for $j = 1, \ldots, r_{l-1}$, since

$$z_i^l = \beta_i^l + w_{i1}^l a_1^{l-1} + \cdots + w_{ij}^l a_j^{l-1} + \cdots + w_{ir_{l-1}}^l a_{r_{l-1}}^{l-1},$$

as usual. This yields lines 9 and 10 of our algorithm.

On the other hand, if we want to connect to the previous layer, we have

$$\frac{\partial L}{\partial a_j^{l-1}} = \frac{\partial L}{\partial z_i^l} \frac{\partial z_i^l}{\partial a_j^{l-1}} = \delta_i^l w_{ij}^l.$$

When we actually compute this quantity, however, we are already thinking about the previous layer. If we therefore shift out perspective, focusing instead on layer $l - 1$, we can make the substitutions[3] $l \to l + 1$, $i \to k$, $j \to i$, we may rewrite this quantity as

$$\frac{\partial L}{\partial a_i^l} = \frac{\partial L}{\partial z_k^{l+1}} \frac{\partial z_i^{l+1}}{\partial a_i^l} = \delta_k^{l+1} w_{ki}^{l+1},$$

which is exactly the summand occurring on line 7 of the algorithm. Note that we are summing over the neurons of the subsequent layer, representing each possible path branch one might take when jumping from neuron i of layer l to layer $l + 1$.

Using the backpropagation algorithm, we can compute the gradient of an arbitrary loss function with respect to the full set of connection weights of the neural network. Using these gradients, we can then deploy stochastic or mini-batch gradient descent to learn the values of the connection weights. In practice, one typically initializes the weights using small random numbers.

Matrix Formulation of Backpropagation

The matrix version of the backpropagation algorithm is shown in Algorithm 13.5. There are a few points to be made regarding notation. First, since the quantity $\phi_l(z^l)$ is defined as the vector quantity $\phi_l(z^l) = \langle \phi_l(z_1^l), \ldots, \phi_l(z_{r_l}^l) \rangle$, its gradient with respect to z^l is the diagonal matrix

$$D\phi_l(z^l) = \begin{bmatrix} \phi_l'(z_1^l) & 0 & \cdots & \vdots \\ 0 & \phi_l'(z_2^l) & \cdots & \vdots \\ \vdots & \ddots & \ddots & \vdots \\ 0 & 0 & \cdots & \phi_l'(z_{r_l}^l) \end{bmatrix}.$$

---

[3] We use i to index the *current* layer l, j to index the previous layer $l - 1$, and k to index the subsequent layer $l + 1$.

Note that the matrix product $D\phi_l(z^l) \cdot x$, for a given $x \in \mathbb{R}^{r_l}$, is equivalent to the componentwise product $\phi'_l(z^l) \odot x$, where we define $\phi'_l(z^l)$ as a vector quantity, similar to the definition of $\phi_l(z^l)$.

Later when we code our matrix-based implementation, we will consider the softmax transformation to be a particular type of activation. In this context, the gradient will necessarily be a matrix.

The second (brief) note is that the gradient $\nabla_{\hat{y}} L$, which we shall interpret as a column vector, is defined in the obvious way.

Finally, on line 8, we use the *outer product* $x \otimes y = xy^T$ for two vectors $x \in \mathbb{R}^p$, $y \in \mathbb{R}^q$, to represent the $p \times q$ matrix with components

$$[x \otimes y]_{ij} = x_i y_j,$$

for $i = 1, \ldots, p$ and $j = 1, \ldots, q$.

Gradient of the Softmax Layer

The quantities δ_i^m for a softmax layer naturally require separate treatment. In Figure 13.4, we may regard the sigma neurons and softmax neurons as cohabitors of a single output layer. We must therefore evaluate the derivative of the loss with respect to the linear combinations z_i^m. Since z_i^m feeds into each softmax neuron, as opposed to its counterpart alone, we must compute the summation

---

Algorithm 13.5: Backpropagation (matrix version).

Input: a training instance (x, y);
a given neural network of size m;
a set of m activation functions $\phi_1, \ldots, \phi_m$;
a given set of (current) bias vectors $\mathcal{B} = \{\beta^1, \ldots, \beta^m\}$;
a given set of (current) weights $\mathcal{W} = \{W^1, \ldots, W^m\}$
Output: The full gradient $\nabla_{\mathcal{W}} L$.

1 Compute $\hat{y} = f(x, \mathcal{W})$ using the feedforward algorithm, saving the values of all intermediary variables

2 **for** *each layer $l = m, \ldots, 1$ (in descending order)* **do**

3   **if** *layer l is output layer $l = m$* **then**

4   Set $\delta^m = D\phi_m(z^m) \cdot \nabla_{\hat{y}} L$

5   **else**

6   Set $\delta^l = D\phi_l(z^l) \cdot (W^{l+1})^T \cdot \delta^{l+1}$

7   **end**

8   Set $\dfrac{\partial L}{\partial W^l} = \delta^l \otimes a^{l-1}$

9   Set $\dfrac{\partial L}{\partial \beta^l} = \delta^l$

10 **end**

11 **return** $\nabla_{\mathcal{W}} L = \left\{ \dfrac{\partial L}{\partial W^l} \right\}_{l=1}^m$ and $\nabla_{\mathcal{B}} L = \left\{ \dfrac{\partial L}{\partial \beta^l} \right\}_{l=1}^m$

$$\delta_i^m = \frac{\partial L}{\partial z_i^m} = \sum_{k=1}^{c} \frac{\partial L}{\partial \hat{p}_k} \frac{\partial \hat{p}_k}{\partial z_i^m}.$$

Now, given our definition of the softmax transform (Equation (13.4)), one easily computes the following partial derivatives

$$\frac{\partial \hat{p}_k}{\partial z_i^m} = \frac{\exp(z_k)\delta_{ik}}{\displaystyle\sum_{j=1}^{c} \exp(z_j)} - \frac{\exp(z_k)\exp(z_i)}{\left(\displaystyle\sum_{j=1}^{c} \exp(z_j)\right)^2} = \hat{p}_k I_{ik} - \hat{p}_i \hat{p}_k,$$

where $I_{ik} = \mathbb{I}[i = k]$ is the identity. We may combine the preceding two equations into the single matrix equation

$$\delta^m = (\mathrm{diag}(\hat{p}) - \hat{p} \otimes \hat{p}) \cdot \nabla_{\hat{y}} L. \tag{13.7}$$

We may make a further simplification if the loss function is log-loss, i.e., cross-entropy (Equation (6.20)):

$$L(y, \hat{p}) = -\sum_{k=1}^{c} y_k \log(\hat{p}_k),$$

where y is the one-hot encoded vector of class labels. For the case of the cross-entropy loss function, we therefore have the result that

$$\delta_i^m = -\sum_{k=1}^{c} \frac{y_k}{\hat{p}_k} \left(\hat{p}_k \delta_{ik} - \hat{p}_i \hat{p}_k\right),$$

which simplifies as

$$\delta_i^m = \hat{p}_i - y_i. \tag{13.8}$$

For multiclass classification problems with a softmax layer, we simply keep calm, replace line 5 of Algorithm 13.4 (or line 4 of Algorithm 13.5) with the preceding equation, and carry on from there.

13.2.3 Deep Learning

Now that we have derived the necessary formulas and algorithms to compute the gradients of our loss function with respect to the full set of connection weights, we may discuss actually implementation of the learning algorithm, which typically comes in flavors of gradient descent. We will briefly touch on a few of these topics; for further details, see Aggarwal [2018]. For additional background specifically for optimization strategies, see Chong and Zak [2008].

Challenges

When building and training artificial neural networks, there are several challenges that can arise and that one should be aware of.

Vanishing and Exploding Gradients

Vanishing and exploding gradients refers to the general phenomenon that gradients have the propensity to exponentially decay (or grow) with the number of layers as one moves backward through a neural network during backpropagation. To see why, consider an illustrative example of a simple neural network with ten layers that each use the logistic activation function and posses only a single neuron. The maximum value of the derivative of the logistic function is only 0.25; recall Figure 13.6. Since these derivatives are multiplied during backpropagation as we move back through the network, even if each neuron is precisely at the maximum value ($z = 0$), the magnitude of the gradient in the first layer is approximately *one millionth* ($1/4^{10}$) of the magnitude of the gradient in the final layer, resulting in a glacially slow parameter movement through the early layers. The hyperbolic tangent function, of course, has a maximum derivative of 1, but like the logistic function, it quickly decays to zero outside of a short band. This is a large part why the ReLU and hard tanh functions have become go-to choices for deep neural architectures.

One issue with the ReLU function, however, is that its gradient is zero for negative values of the input, resulting in "dead neurons." This can function as a tuning of sorts, in which unnecessary neurons are naturally removed. This may be viewed as a natural form of neural pruning, which can be advantageous given the limits to precisely tuning the number of neurons in each layer. However, one must be on guard that too many are not destroyed resulting in ineffective models. Slow learning rates tend to offset this problem, as does the replacement with the ReLU function with a *leaky ReLU*, which is defined by

$$\phi(z) = \begin{cases} \alpha z & \text{for } z \leq 0 \\ z & \text{for } z > 0 \end{cases},$$

for a given parameter $\alpha \in (0, 1)$. This allows backpropagation to continue to propagate information backwards through dead neurons at a reduced rate α.

Initialization of Weights

In practice, we begin our stochastic gradient descent with a random initialization of the weights, say $W_{ij}^l \sim N(0, \epsilon^2)$, for some small $\epsilon << 1$ (e.g., $\epsilon = 0.01$). This approach, however, can still result in some instability, for neural architectures in which individual layers have a vastly different number of neurons between them, since the variance of the outputs scales linearly with the number of inputs. It is therefore common in practice to initialize weights following

$$\beta_i^l = 0 \tag{13.9}$$
$$W_{ij}^l \sim N\left(0, r_{l-1}^{-1}\right); \tag{13.10}$$

i.e., we set the weights of the bias neurons to zero and the non-bias neurons to be random values with a standard deviation of $1/\sqrt{r_{l-1}}$.

Hyperparameter Tuning

For a given problem, not only do we need to determine the connection weights, but we must also tune various hyperparameters, including

1. the number of hidden layers m;
2. the number r_l of neurons in each layer;
3. the learning rate η (discussed below).

In order to choose our hyperparameters, we can use train–test–validate sets, as was discussed in Section 6.1.5. This still leaves us with the problem of *which* combinations of hyperparameters to test.

A common approach, known as *grid search*, involves discretizing the hyperparameters and training the model over all combination of values. The number of grid points, however, increases exponentially with the number of hyperparameters, as it entails creating a discretization of an equally high-dimensional hypercube: a grid with h hyperparameters, each with n values, grid search would entail training the model over n^h combinations of hyperparameters.

As an alternative to grid search, we could uniformly sample each hyperparameter within its respective range, as proposed by Bergstra and Bengio [2012]. This can be coupled with multi-resolution sampling, in which the full range of hyperparameters is sampled on the first pass, and geometrically smaller range is explored on subsequent passes, each centered around the optimal values of the prior pass.

Parameters that scale across a range of orders of magnitude, like learning rate or regularization rate (discussed in our next paragraph), can of course be sampled logarithmically instead of uniformly. For instance, if we are trying to find an optimal learning rate between 0.001 and 0.1, we could try 0.001, 0.01, and 0.1, as opposed to a linear division of the range $[0.001, 0.1]$.

Regularization

Given the large number of parameters, neural networks are prone to overfitting on the training set. In order to guard against overfitting, we can deploy ℓ_1 (lasso) or ℓ_2 (ridge) regularization (Definition 7.4) by considering instead the augmented loss functions

$$L_1(y, \hat{y}) = L(y, \hat{y}) + \lambda \sum_{l=1}^{m} \sum_{i=1}^{r_l} \sum_{j=1}^{r_{l-1}} \left| w_{ij}^l \right|, \tag{13.11}$$

$$L_2(y, \hat{y}) = L(y, \hat{y}) + \lambda \sum_{l=1}^{m} \sum_{i=1}^{r_l} \sum_{j=1}^{r_{l-1}} \left(w_{ij}^l \right)^2, \tag{13.12}$$

respectively, where λ is a tunable hyperparameter. Note that the bias weights β_i^l are not penalized. Naturally, it is straightforward to compute the gradient of the regularization terms with respect to the weights, adding the result to the gradient computed during backpropagation.

ℓ_2 regularization is most commonly used in practice, as it usually results in models with higher accuracy. As we saw in Chapter 6, however, ℓ_1 regularization is also useful as it often results in a *sparse* set of parameters, resulting in simpler models that are faster to train. This strategy can also be regarded as a way to prune large networks.

Descent into Madness

In this text, we will focus on stochastic and mini-batch gradient descent for simplicity. These algorithms are provided in Algorithm 13.6; but see also Algorithm 5.2 and Definition 5.5 for our earlier discussion. However, there are many variations to simple gradient descent that can greatly improve performance:

1. *momentum methods* attempt to correct the "zigzagging" effect observed in gradient descent by updating the gradient in the direction of a "momentum" vector;
2. *learning rate decay* is a strategy coupled with standard gradient descent in which the learning rate decays exponentially (or inversely) with the number of training steps;
3. *spectrum learning rates* is a collection of strategies (AdaGrad, RMSProp, RMSProp with Nesterov momentum, AdaDelta, Adam) that allow for each parameter to have its own learning rate, thereby allowing the learning algorithm to speed up in a targeted fashion in the areas that need it most;
4. *Netwon's method* involves using the Hessian of the loss function, which entails computing the second derivatives of the loss function over the full set of weights;
5. *conjugate gradient method*, also referred to as "Hessian-free optimization" can solve a c-dimensional quadratic function in a total of c steps by ensuring that each update is "orthogonal" in an appropriate sense to the previous set of updates.

For an overview of each of these techniques in relation to deep learning, see Aggarwal [2018].

13.3 Python Implementation

In this section we will build and train our own neural networks from scratch. We will follow two different strategies. The first is a neuron-based approach, which most directly models the biological brain. This approach is used primarily for its illustrative value. The second approach is a layer-based, or vector-based, approach, which leverages the speed of numpy's built-in matrix operations for improved performance. We then train a model using mini-batch gradient descent. Finally, we discuss how to build and train neural networks using TensorFlow 2.0 in Python.

Algorithm 13.6: Stochastic ($b = 1$) and mini-batch ($b > 1$) gradient descent for deep learning.

Input: training data $\mathcal{D} = \{(x_i, y_i)\}_{i=1}^{n}$;
  a given neural network of size m;
  a set of m activation functions $\phi_1, \ldots, \phi_m$;
  learning rate η;
  batch size b (typically a power of 2)

Output: Set of weights $\mathcal{W} = \{W^l\}_{l=1}^{m}$ and $\mathcal{B} = \{\beta^l\}_{l=1}^{m}$.

1 Set $\beta_i^l = 0$ for $l = 1, \ldots, m$ and $i = 1, \ldots, r_l$
2 Set $w_{ij}^l \sim \mathrm{N}(0, 1/r_{l-1})$ for $l = 1, \ldots, m$, $i = 1, \ldots, r_l$, and $j = 1, \ldots, r_{l-1}$
3 Set $\mathcal{W} = \{w_{ij}^l\}_{l,i,j}$
4 Set $\mathcal{B} = \{\beta_i^l\}_{l,i}$
5 Set $converged = $ **False**
6 **while not** $converged$ **do**
7 $\quad$ Compute sample index $I_\mathcal{B} = $ getRandomIndex($|\mathcal{D}|, b$)
8 $\quad$ **for** $s \in I_\mathcal{B}$ **do**
9 $\quad\quad$ Compute $\hat{y}_s = f(x_s, \mathcal{W}, \mathcal{B})$ using the feedforward algorithm
10 $\quad\quad$ Compute $\nabla_\mathcal{W} L_s = \left\{ \dfrac{\partial L}{\partial w_{ij}^l}(y_s, \hat{y}_s) \right\}_{l,i,j}$ using backpropagation
11 $\quad\quad$ Compute $\nabla_\mathcal{B} L_s = \left\{ \dfrac{\partial L}{\partial \beta_i^l}(y_s, \hat{y}_s) \right\}_{l,i}$ using backpropagation
12 $\quad$ **end**
13 $\quad$ Update $\mathcal{W} = \mathcal{W} - \eta \sum_{s \in I_\mathcal{B}} \nabla_\mathcal{W} L_s$
14 $\quad$ Update $\mathcal{B} = \mathcal{B} - \eta \sum_{s \in I_\mathcal{B}} \nabla_\mathcal{B} L_s$
15 $\quad$ $converged = $ checkConverged($\mathcal{D}, \mathcal{W}, \mathcal{B}$)
16 **end**
17 **return** $\mathcal{W}$ and $\mathcal{B}$

13.3.1 Activation Functions

To begin, we need to code up some activation functions. An abstract
`Activation` class is provided in Code Block 13.1. Notice the usage of the
static method _vectorize, which will serve as a wrapper to the function
call f and its derivative df in each subclass.

```python
class Activation(ABC):

    @staticmethod
    def _vectorize(func):
        def wrapper(*args, **kwargs):
            vfunc = np.vectorize(func, otypes=[float])
            return vfunc(*args, **kwargs)
        return wrapper

    @abstractmethod
    def f(self, x):
        pass

    @abstractmethod
    def df(self, x):
        pass
```

Code Block 13.1: `Activation` abstract class.

Note that the `numpy.vectorize` method is primarily written for convenience, not performance; i.e., it is not a true *vectorization* of one's code, as
it does not leverage parallelization or multi-threading, etc. Rather, it has
the performance of an ordinary *for*-loop. Additional activation functions
can be encoded following our cookie-cutter template of Code Block 13.2.

```python
class ActivationReLU(Activation):

    @Activation._vectorize
    def f(self, x):
        return x if x > 0 else 0

    @Activation._vectorize
    def df(self, x):
        return 1 if x > 0 else 0
```

Code Block 13.2: `ActivationReLU`: ReLU activation function.

13.3.2 Neuron-based (Scalar) Implementation

A natural first approach at coding artificial neural networks is as a graph of individual neurons, that are connected together in a systematic way. We may think of this as a *neuron-first* approach. It suffers, however, from the drawback that it fails to vectorize the operations involved in deep learning. We therefore regard this exercise as more illustrative in nature, designed to aid the conceptual understanding on the inner workings of complex networks.

Neurons

We begin with our code for a single artificial neuron, as shown in Code Block 13.3. We will diverge slightly from our mathematical notation developed in the preceding sections. Since each neuron has its own bias, we track the bias of a neuron separately from the weights, initializing the bias to zero. We initialize our weights following the strategy of Equation (13.10).

Other than the value of a neuron's bias, we must also consider the number of input connections, the activation function, and whether or not the bias should be used in the actual computation. We therefore require these variables to be passed into the __init__ method of the `Neuron` class on line 3. We will attach the weights of the input connects to the receiving neuron, using the `weights` attribute. We will also need to track the "state" of each neuron, represented by the previous values of the sum z_i^l, the output value a_i^l, the gradient of the weights $\nabla_{W^l} L$ and bias $\partial L/\partial \beta_i^l$, and the value of δ_i^l used during backpropagation. We initialize these values to zero on lines 11–16. Since these variables are all attached to the ith neuron of layer l (something we will construct *outside* of the `Neuron` class), we do not need to enumerate them within the class itself. In other words, each `Neuron` object should be agnostic to its place in the world.

Next, we need to consider the types of operations a neuron must undergo, and how we should implement their corresponding methods. There are essentially two primary operations: feedforward computation and backpropagation. For the former, we include the `activate` method, which takes an input vector and computes the given neuron's output $a = \phi \circ \omega(x)$. We will save the values of the input value (as `last_value`), the intermediary variable $z_i^l = \sum_{j=1}^{r_{l-1}} w_{ij}^l a_j^{l-1}$, and the activation value as attributes, as these are each used during the backpropagation phase.

For the `backpropagate` method, we will assume an input value that is the current neuron's component of the vector joining the forward weights and forward delta-values into a single quantity, which we call *weighted forward deltas*. We will discuss this in more depth momentarily (in our next paragraph on *Neuron Layers*. For now, suffice it to say that this quantity represents the parenthesized quantity on line 7 of Algorithm 13.4, which nicely captures everything we need to know about *the future* during our

```
 1  class Neuron:
 2
 3      def __init__(self, phi=None, n_inputs=1, bias=0, use_bias=True,
            lambda_=0.1, penalty=None):
 4          self.phi = phi
 5          self.bias = bias
 6          self.use_bias = use_bias
 7          self.lambda_ = lambda_
 8          self.penalty = penalty
 9          self.n_inputs = n_inputs
10          self.weights = np.random.normal(0, scale=1/n_inputs,
                size=n_inputs)
11          self.z = 0
12          self.value = 0
13          self.dW = np.zeros(n_inputs)
14          self.dBias = 0
15          self.delta = 0
16
17      def activate(self, x):
18          self.last_input = x
19          self.z = self.bias + np.dot(x, self.weights)
20          self.value = self.phi.f(self.z)
21          return self.value
22
23      def backpropagate(self, weighted_forward_delta):
24          self.delta = weighted_forward_delta * self.phi.df(self.z)
25          self.dBias = self.delta if self.use_bias else 0
26          self.dW = self.delta * self.last_input
27          return self.delta * self.weights
28
29      def step(self, eta):
30        if self.use_bias:
31              self.bias -= eta * self.dBias
32          self.weights -= eta * self.dW
```

Code Block 13.3: Neuron class

backpropagation step. From there, it is straightforward to compute the current neuron's δ_i^l value, and the gradient with respect to the given neuron's bias and input weights.

Finally, we implement the gradient "step" in our final method, step, which actually updates the neuron's bias (if applicable) and weights.

Neural Layers

We next discuss how neurons can be bundled into the concept of a neural layer. We will save ourselves some trouble later on, when we replace our neuron-based implementation with a matrix-based one, if we take a step back now and think about the neural layer abstractly and ask what we need to achieve from it. In this regard, the functionality of a layer is not dissimilar to the functionality of an individual neuron: we need to be able to run both feedforward and backpropagation passes through individual layers. The neural layer, however, has added structure as a collection of neurons and must therefore be able to handle the higher-order orchestration of the feedforward and backpropagation operations.

Let us consider again the neural network shown in Figure 13.7, with a focus on layer $l = 3$. Abstracting the concept of a neural layer, we can consider the basic functionality as shown in Figure 13.13. The feedforward

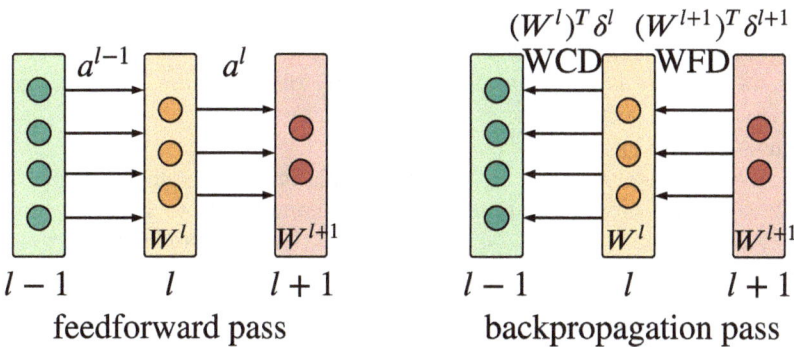

Fig. 13.13: Schematic of the feedforward (left) and backpropagation (right) passes of an abstract neural layer.

pass is relatively straightforward: the previous output is passed as an input into the current layer, which is then transformed into a new output for the following layer.

In order the better understand the input/output requirements for backpropagation, let us consider again Algorithms 13.4 and 13.5. The information that comes from the following layer is given as the summation in line 7 of Algorithm 13.4, or its equivalent matrix representation in Algorithm 13.5. We will refer to this quantity as the *weighted forward deltas (WFD)*, which is defined as the vector

$$WFD = (W^{l+1})^T \delta^{l+1} = \left\langle \sum_{k=1}^{r_{l+1}} w_{ki}^{l+1} \delta_k^{l+1} \right\rangle_{i=1}^{r_l}.$$

For the final output layer, we can replace this with the gradient of the loss function. Each neuron in our neuron-based implementation is then to receive only a single component of the vector of weighted forward deltas as an input to that neuron's `backpropagate` method. If this is the required input from the next layer during backpropagation, then this quantity must also be the required output of our current layer, in order that the algorithm continues to cascade backwards through the process of backpropagation. We will refer to a layer's backpropagation output as the *weighted current deltas (WCD)*, obtained by replacing $l + 1 \rightarrow l$ in the above equation.

Finally, note that the weights are always attached to their given layers. In the neuron-based implementation, they are embedded directly into the individual neurons that comprise the layer. In the matrix-based implementation, we will forgo the use of individual neurons and replace it with a matrix representation of each neural layer's inner workings. We now have all the requirements we need to abstract the concept of a neural layer into an abstract class.

The neuron-based implementation of a neural layer is provided by the `NeuronLayer` class of Code Block 13.4. Note that the `Neuron.backpropagate` method of Code Block 13.3 already returns the product of that neuron's δ and its weights, so the summation on line 7 of Algorithm 13.4 need only sum over these vector outputs. The result is passed as an output, which is then consumed by the previous layer.

Since the neuron-based implementation is mostly for illumination, and since softmax layers are more readily implemented from a matrix-based approach, we will postpone our implementation of a softmax layer until later.

Layer Prescription and Loss

Now that we have the workings of the neural layers, we need a brain to orchestrate the operations across the neural net. We will accomplish this by writing a `NeuralNetwork` class. In constructing such a class, however, we will need to specify the neural architecture. This will consist of a prescription for the overall number of inputs and outputs, a loss function, and the design of the hidden layers, with each hidden layer specified by the triple consisting of the number of neurons, the activation function, and whether or not bias is allowed. Therefore, before we build our `NeuralNetwork` class, we will need to determine how we will pass our architecture and loss function into the constructor.

In order to specify the triple (number of neurons, activation function, and whether to use a bias) that defines a neural layer, we will subclass the built-in `namedtuple` class, which can be used as a factory that dynamically defines a new class on the fly with a prescribed set of attributes and their corresponding defaults. For construction, see Code Block 13.5.

```
class NeuronLayer:

    def __init__(self, phi=None, n_inputs=1, n_outputs=1,
            use_bias=True, lambda_=0.1, penalty=None):
        self.phi = phi
        self.use_bias = use_bias
        self.lambda_ = lambda_
        self.penalty = penalty
        self.n_inputs = n_inputs
        self.n_outputs = n_outputs
        self.neurons = [Neuron(phi=phi, n_inputs=n_inputs,
                use_bias=use_bias) for i in range(n_outputs)]

    def __len__(self):
        return len(self.neurons)

    def activate(self, x):
        return np.array([n.activate(x) for n in self.neurons])

    def backpropagate(self, weighted_forward_deltas):
        weighted_current_deltas = np.zeros(self.n_inputs)
        for neuron, weighted_forward_delta in zip(self.neurons,
                weighted_forward_deltas):
            weighted_current_deltas += \
                    neuron.backpropagate(weighted_forward_delta)
        return weighted_current_deltas

    def step(self, eta):
        for neuron in self.neurons:
            neuron.step(eta)
```

Code Block 13.4: NeuronLayer class

```
from collections import namedtuple
LayerDefinition = namedtuple('LayerDefinition', ['size', 'phi',
    'use_bias', 'lambda_', 'penalty'], defaults=[1,
    ActivationReLU(), True, 0.1, None])
layer_1 = LayerDefinition(size=5, phi=ActivationReLU(),
    use_bias=False)
layer_2 = LayerDefinition(size=10, phi=ActivationTanh())
layer_defs = [layer_1, layer_2]
```

Code Block 13.5: LayerDefinition class and example construction.

Similarly, we can encode our loss function as shown in Code Block 13.6. Constructing an object from the `Loss` class is a bit like a factory method, in that the type of loss function is determined by the `loss_type` keyword. There are more elegant ways one might encode a loss function, but for our purposes, using simple `if` statements suffices.

```python
class Loss:

    def __init__(self, loss_type):
        self.type = loss_type

    def f(self, y, y_hat):
        if self.type == 'squared_error':
            return (y - y_hat)**2
        if self.type == 'log':
            return -np.sum(y * np.log(y_hat), axis=0)

    def df(self, y, y_hat):
        if self.type == 'squared_error':
            return -2 * (y - y_hat)
        if self.type == 'log':
            return -y / y_hat
```

Code Block 13.6: `Loss` class; implementation of squared-error and log loss.

Neural Networks

Now we are ready to build our network. The constructor will require a specification of the number of inputs, an array of `LayerDefinition` objects, which specifies the neural architecture, and a loss function. The code is shown in Code Block 13.7. Note that we use `MatrixLayer`, which is provided later in Code Block 13.8. We could equivalently swap this out with `NeuronLayer` to the same effect; since both classes follow the same API, they may be used interchangeably. One could further write a simple factory method to control what type of layer is used, but since our ultimate goal is to use the matrix-version of a neural network, we find this step to be unnecessary.

Dimensionality of Inputs and Outputs

By convention, the *rows* of the feature matrix and target vector correspond to individual training instances, whereas the *columns* correspond to the individual features or labels, respectively. This is antithetical to our mathematical development, where we interpret vectors as column vectors, not

```
 1  class NeuralNetwork(Model):
 2
 3      def __init__(self, n_inputs=1, layer_defs=[], loss_type=None,
             batch_size=1, eta=0.1):
 4          self.batch_size = batch_size # size > 1 invalid
 5          self.eta = eta #learning rate
 6          self.n_inputs = n_inputs
 7          self.loss = Loss(loss_type)
 8          self.layers = []
 9          for ldef in layer_defs:
10              self.layers.append(MatrixLayer(phi=ldef.phi,
                    n_inputs=n_inputs, n_outputs=ldef.size,
                    use_bias=ldef.use_bias, lambda_=ldef.lambda_,
                    penalty=ldef.penalty))
11              n_inputs = ldef.size
12
13      def feedforward(self, X):
14          for layer in self.layers:
15              X = layer.activate(X)
16          return X
17
18      def backpropagate(self, y, y_hat, eta=0.1):
19          weighted_deltas = self.loss.df(y, y_hat)
20          for layer in reversed(self.layers):
21              weighted_deltas = layer.backpropagate(weighted_deltas)
22              layer.step(eta)
23
24      def train_batch(self, X, y):
25          y_hat = self.feedforward(X)
26          self.backpropagate(y, y_hat, eta=self.eta)
27
28      def train(self, X, y, n_rounds=10):
29          for _ in range(n_rounds):
30              idx = np.random.choice(len(X), size=self.batch_size,
                    replace=False)
31              X_train = X[idx]
32              y_train = y[idx]
33              self.train_batch(X_train.T, y_train.T)
34
35      def predict(self, X):
36          return self.feedforward(X.T).T
```

Code Block 13.7: NeuralNetwork class.

row vectors, in order to be consistent with linear algebra, where it is conventional to write the matrix–vector product as $A \cdot x$ (where x is a column vector), not $r \cdot A$ (where r is a row vector). This is, coincidentally, why our definition of the weight matrix is the transpose of many other author's definitions, who, presumably, were primarily in the mindset to define the weight matrix consistent with the inputs and outputs of a training algorithm; i.e., from a machine-learning perspective.

To resolve this, we leverage the *abstraction* principle of object-oriented design: *the operations of an object can be defined without reference to its internal implementation*. Thus, our `NeuralNetwork` class will use an API for its `train` and `predict` methods that is consistent with our previous `Model` class (as well as the numerous models within *sci-kit learn*), while leveraging the proper linear-algebra convention during its implementation. We feel this approach is more natural, so that our programmatic implementation is more analogous with its underlying maths. Moreover, this is easily achieved by simply applying a transpose operator to our mini-batches, prior to passing them into the `train_batch` method, as shown in Code Block 13.7.

Visually, we may think of the external API form of the feature matrix and target vector with dimensionality as shown in Figure 13.14. Upon transposing these matrices, we have the internal math form shown in Figure 13.15.

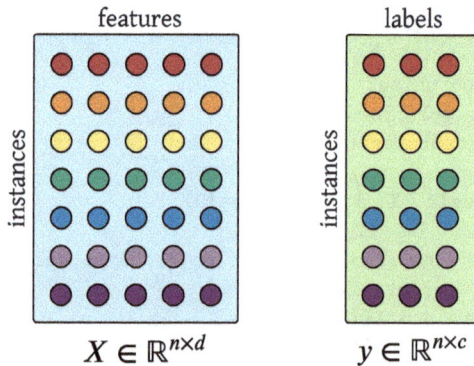

Fig. 13.14: Dimensionality of feature matrix and target vector: external I/O form. Each instance is a separate color.

As a rule of thumb, everything deeper than the `train` and `predict` methods will assume its inputs and outputs are of the internal math form (Figure 13.15). When training mini-batches, the columns of all the internal variables will therefore represent individual instances from the mini-batch. This works seamlessly with standard linear algebra, since, given feature

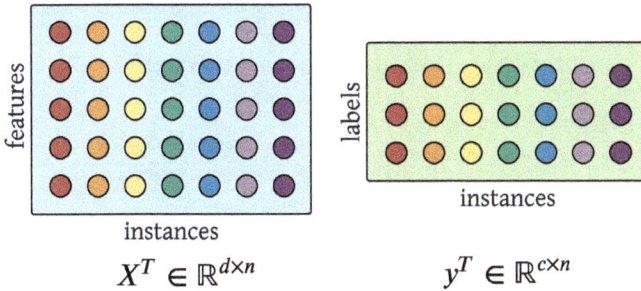

Fig. 13.15: Dimensionality of feature matrix and target vector: internal math form. Each instance is a separate color.

vectors $x_1, \ldots, x_b$, the matrix product $Z = W \cdot X^T$ yields

$$Z = W \cdot \begin{bmatrix} \top & \top & & \top \\ x_1 & x_2 & \cdots & x_b \\ \bot & \bot & & \bot \end{bmatrix} = \begin{bmatrix} \top & \top & & \top \\ W \cdot x_1 & W \cdot x_2 & \cdots & W \cdot x_b \\ \bot & \bot & & \bot \end{bmatrix}.$$

13.3.3 Layer-based (Matrix) Implementation

When creating multiple classes that can be used interchangeably, it is best to define an abstract parent class, so that the inputs and outputs can be enforced. We leave this as an exercise to the reader (see Exercise 13.8). As promised the MatrixLayer class is implemented below, in Code Block 13.8, which can replace our existing NeuronLayer class from Code Block 13.4.

Finally, we can implement softmax layers by creating a custom vectorized softmax activation class, as shown in Code Block 13.9. This way, we may prescribe a softmax layer within the LayerDefinition named tuple. Notice that the methods of the ActivationSoftmax class are vector operations that capture interdependencies among the inputs, which is why the vectorize wrapper is not used. This is also why it is more natural to implement the softmax layer in the context of a matrix approach, as the inputs and outputs of each layer are captured as vectors, as opposed to breaking them down and passing them into individual neurons.

We must further take care to ensure that both a typical and softmax activation function may be used interchangeably within the MatrixLayer class; in particular, in the backpropagate method. Of concern is one activation returning a vector for df and another returning a matrix (i.e., a two-dimensional array). We solve this by adding a semiprivate _diag method, which diagonalizes a one-dimensional array, while leaving a two-dimensional array unchanged.

```
1   class MatrixLayer:
2
3       def __init__(self, phi=None, n_inputs=1, n_outputs=1,
            use_bias=True, lambda_=0.1, penalty=None):
4           self.phi = phi
5           self.use_bias = use_bias
6           self.lambda_ = lambda_
7           self.penalty = penalty
8           self.n_inputs = n_inputs
9           self.n_outputs = n_outputs
10          self.bias = np.zeros(n_outputs)
11          self.weights = np.random.normal(0, scale=1/n_inputs,
                size=(n_outputs, n_inputs))
12          self.grad_bias = np.zeros(n_outputs)
13          self.grad_weights = np.zeros((n_outputs, n_inputs))
14          self.z = np.zeros(n_outputs)
15          self.values = np.zeros(n_outputs)
16          self.deltas = np.zeros(n_outputs)
17
18      def __len__(self):
19          return self.n_outputs
20
21      def _diag(self, x):
22          if x.ndim == 1:
23              return np.diag(x)
24          if x.shape[1] == 1:
25              return np.diag(x.reshape(-1))
26          return x
27
28      def activate(self, x):
29          self.last_input = x
30          self.z = self.bias + self.weights @ x
31          self.value = self.phi.f(self.z)
32          return self.value
33
34      def backpropagate(self, weighted_forward_deltas):
35          ### NOT VALID FOR BATCH ###
36          self.delta = self._diag(self.phi.df(self.z)) @
                weighted_forward_deltas
37          self.grad_weights = np.outer(self.delta, self.last_input)
38          self.grad_bias = self.delta
39          weighted_current_deltas = self.weights.T @ self.delta
40          return weighted_current_deltas
41
42      def step(self, eta):
43          if self.use_bias:
44              self.bias -= eta * self.grad_bias
45          self.weights -= eta * self.grad_weights
```

Code Block 13.8: MatrixLayer class.

```
1    class ActivationSoftmax(Activation):
2
3        def f(self, x):
4            return np.exp(x) / np.exp(x).sum(axis=0)
5
6        def df(self, x):
7            p_hat = self.f(x)
8            return np.diag(p_hat) - np.outer(p_hat, p_hat)
```

Code Block 13.9: ActivationSoftmax class.

In addition, we note that the code as written does not allow for batch gradient descent. This primarily breaks down in the backpropagate method, which would require additional complexity as all two-dimensional arrays would need to be rewritten as tensors. We will content ourselves at present with stochastic gradient descent and leave the generalization as a project for the interested reader.

13.4 Advanced Neural Architectures

In this section, we will introduce some of the more advanced neural architectures that underlie applications such as image recognition and speech processing. For more details, see the usual references Aggarwal [2018], Gèron [2019], and Goodfellow, *et al.* [2016].

13.4.1 Autoencoders

Our first, and simplest, example of an advanced structure is that of an *autoencoder*. From the outside, autoencoders appear strange in that not only do their output nodes mirror their input nodes, but also the input vectors are used as labels during training; i.e., they learn how to predict the input from the input. Thus, from a standpoint of input and output alone, they appear most puzzling. The key to understanding them, however, is their internal workings. Autoencoders consist of a number of hidden layers of smaller dimensionality than their input and output, for the purpose of creating a low-dimensional representation of the data.

A simple example of an autoencoder is shown in Figure 13.16. It is typical in practice for autoencoders to consist of an odd number of layers, and for the number of neurons to be symmetric around the central layer. In general, both the encoder half and decoder half of an autoencoder will be multilayer neural networks. Once trained, an autoencoder will consist of two functioning components—an encoder and a decoder—that can be used independently. The encoder can create a low-dimensional encoding of

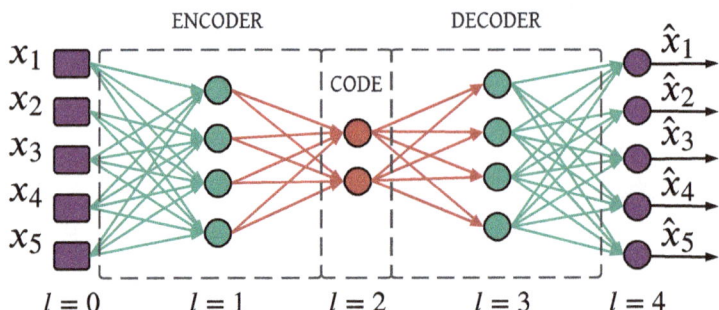

Fig. 13.16: A simple autoencoder.

high-dimensional feature vectors, which can then be used in training many models. Thus, an encoder is not unlike principal component analysis and other feature subspace selection techniques used to reduce dimensionality of complex problems.

13.4.2 Convolutional Neural Networks

Convolutional neural networks are primarily used for image recognition and visual learning tasks. They incorporate a specialized design that mimics the workings of the visual cortex in the brain. We have, of course, seen an example of visual learning before: the NIST handwritten digit problem discussed in Example 7.4. Here, the image consisted of an 8×8 array of pixels, which was flattened into a 64-dimensional feature vector. Such an approach is, however, inefficient and wholly infeasible, when one considers that a typical 4K image consists of nearly nine million pixels. Let's say that our first hidden layer consists of only 1 million pixels, representing a 10:1 loss, this would still yield nearly *nine trillion connections* between the input layer and the first hidden layer. Not only does this represent an explosion in the number of parameters, but also it fails to take advantage of the inherit array structure of the input image: by flattening a two-dimensional structure into a one-dimensional array, we are missing the opportunity to use spatial awareness. Two pixels that are next to each other in the image could end up separated by thousands of rows in a flattened representation of the image. Moreover, color images should properly be considered as three-dimensional objects, as a third dimension is required to encode the color space (e.g., RGB). Finally is the problem of translation and rotation invariance: a banana is a banana if it is in the bottom of the image or the top or if it's up-side-down or right-side-left or if it's peeled or unpeeled or in pajamas or not.

To remedy these issues, we employ a structure called a *convolution layer*, which consists of a small set of weights that form an array called a *filter* or a *kernel*, which can attach to any point in the input array and apply a consistent transformation. This captures two separate principles: weight sharing, in which weights are shared across a variety of neurons, and sparse weights, in which weights are trained only for a subset of connections from one layer to the next.

To make this specific, consider a neural network with input image characterized by a three-dimensional array of size $d_w \times d_h \times d_d$, where d_w represents the image width, d_h the height, and d_d the "color depth." For example, a 64×64 image with RGB color encoding would be represented as a $64 \times 64 \times 3$ array. In general, we can say that the lth layer is of size $r_w^l \times r_h^l \times r_d^l$.

If we were to consider *all possible connections* between one layer and the next, our weight matrix would be a massive matrix with $r_w^{l-1} r_h^{l-1} r_d^{l-1} r_w^l r_h^l r_d^l$ individual weight components. Instead, a convolution layer leverages a much smaller structure called a *filter*, which is a square array (plus depth) of size $f_l \times f_l \times d_l$. The parameter f_l is typically a small, odd number; common values are $f_l = 3$ and $f_l = 5$. Thus, for a three-dimensional color encoding, we are only concerned with learning the $f_l^2 d_l$ parameters of the filter; i.e., 27 or 75 for the case of $f_l = 3, 5$ and $d_l = 3$.

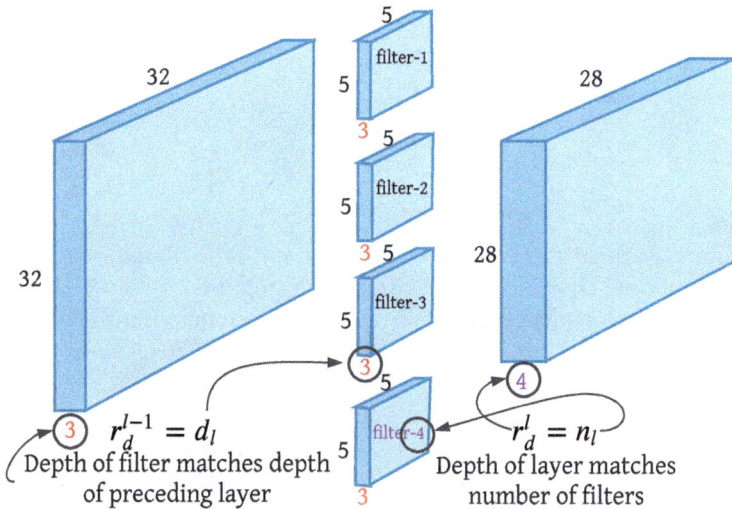

Fig. 13.17: Dimensionality of filters and layers. Each filter contributes one "slice" to the depth of the output layer. The dimensionality loss of the image is typical when not using zero padding.

The first filter $l = 1$ is constrained by the color depth of the input image, so that $d_1 = d_d$. Next, we can allow for *multiple filters* for any given layer. Here, the constraint is that r_d^l is equal to the number of filters for layer l; i.e., each filter adds one unit of depth to the following layer. Finally, the depth of an lth layer filter should equal the depth of its input from the preceding layer; i.e., $d_l = r_d^{l-1}$. These constraints are illustrated in Figure 13.17.

Now that we know what filters look like, let's dive into how they operate. Without further ado, we define the convolution operation as follows.

Definition 13.5. *Given a filter $F \in \mathbb{R}^{f \times f \times d}$ and a three-dimensional input array $A \in \mathbb{R}^{w \times h \times d}$, the unpadded convolution $F \otimes_u A$ is the two-dimensional array*

$$[F \otimes_u A]_{ij} = \sum_{r=0}^{f-1}\sum_{s=0}^{f-1}\sum_{t=0}^{d-1} F_{rst} A_{i+r,j+s,t}, \tag{13.13}$$

defined for $i = 0, \ldots, w - f$ and $j = 0, \ldots, h - f$.

Similarly, the zero-padded convolution *$F \otimes A$ is the two-dimensional array*

$$[F \otimes A]_{ij} = \sum_{r=-m_f}^{m_f}\sum_{s=-m_f}^{m_f}\sum_{t=0}^{d-1} F_{m_f+r,m_f+s,t} A_{i+r,j+s,t}, \tag{13.14}$$

defined for $i = 0, \ldots, w - 1$ and $j = 0, \ldots, h - 1$; where $m_f = (f - 1)/2$ when f is odd; and where zeros are used whenever $A_{i+r,j+s,t}$ is undefined (i.e., zero padding).

Equation (13.14) is equivalent to

$$[F \otimes A]_{ij} = \sum_{r=-0}^{f-1}\sum_{s=-0}^{f-1}\sum_{t=0}^{d-1} F_{rst} A_{i+r-m_f,j+s-m_f,t}.$$

Without zero padding, some information around the boundaries is lost via the convolution operation, leading to underrepresentation of the border pixels in the subsequent layer. This is why it is common in practice to use zero padding, which preserves the spatial footprint from one layer to the next.

Note that in the case of zero padding, the central component of the filter is given by (m_f, m_f), since $m_f = (f - 1)/2$ (assuming f is odd, as is customary). Thus, the center of the filter is attached to each element in the input array. Equation (13.14) can be further interpreted as a moving, limited inner product between the filter and the input array.

This operation is best illustrated by means of an example. We can apply a simple 3×3 filter to a 7×7 input matrix, as shown in Code Block 13.10. This example uses the zero-padded form of convolution, given by Equation (13.14).

```python
F = np.array(
    [[2, 0, 1],
     [0, 1, 0],
     [0, 1, 2]])
A = np.array(
    [[4, 2, 4, 5, 0, 1, 6],
     [7, 2, 8, 8, 0, 4, 0],
     [7, 7, 2, 0, 9, 8, 3],
     [9, 1, 7, 4, 6, 0, 2],
     [2, 0, 4, 2, 2, 6, 4],
     [6, 7, 4, 2, 3, 6, 9],
     [9, 9, 9, 0, 6, 7, 2]])
mf = 1
B = np.zeros((7,7))
for i in range(7):
    for j in range(7):
        for r in range(-mf, mf+1):
            for s in range(-mf, mf+1):
                if (i+r<0) or (i+r>6) or (j+s<0) or (j+s>6):
                    continue
                B[i, j] += F[mf+r, mf+s] * A[i+r, j+s]
```

Code Block 13.10: A simple example of a convolution with zero padding.

This operation is depicted graphically in Figure 13.18. Notice that the (i,j)th component of the output array centers the filter at the (i,j)th component of the input array, and inputs of zero are taken whenever the filter falls off the edge of the grid (i.e., zero padding).

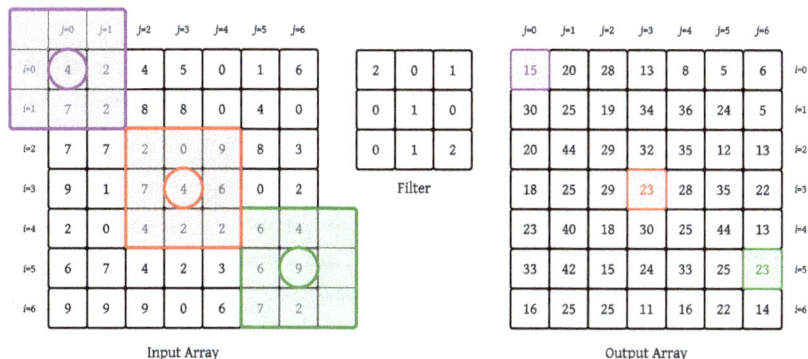

Fig. 13.18: Example of a convolution operation.

Thus, convolution essentially provides a similar operation "scanning" across the image. In this way, a convolutional neural network scans for high-level features of an image, such as lines or boundaries of objects, before refining down to higher-level concepts, like object, structure, animal, etc.

Next, let us consider an example where two filters are applied to an actual RGB image of size 427×640. The code is shown in Code Block 13.11.

```
from sklearn.datasets import load_sample_images
dataset = load_sample_images()
im = dataset.images[0]
FH, FV = np.zeros((7,7,3)), np.zeros((7,7,3))
FH[3,:, :], FV[:, 3, :] = 1, 1
def convolve(A, F):
    w, h, d = A.shape
    mf = int((len(F)-1)/2)
    B = np.zeros((w, h))
    for i in range(w):
        for j in range(h):
            for k in range(d):
                for r in range(-mf, mf+1):
                    for s in range(-mf, mf+1):
                        if (i+r<0) or (i+r>=w) or (j+s<0) or (j+s>=h):
                            continue
                        B[i, j] += F[mf+r, mf+s, k] * A[i+r, j+s, k]
    return B
BH = convolve(im, FH)
BV = convolve(im, FV)
plt.imshow(im); plt.imshow(BH); plt.imshow(BV)
```

Code Block 13.11: Horizontal and vertical filters applied to an RGB image.

Our two filters represent a horizontal and a vertical filter, with zeros throughout a 7×7 array, with the exception of ones placed down the central row and column, respectively. The original image with the two filtered outputs is shown in Figure 13.19. Notice how the horizontal filter accentuates the horizontal lines, while blurring the vertical lines, whereas the vertical filter does the opposite. We call the output of a given filter a *feature map*, as it represents the areas in the image that the filter activated the most. This is why it is common to have multiple filters acting within a given layer. Naturally, the filter weights are learned during backpropagation and not specified in advanced.

Now that we have seen an example of how a simple convolution works, let's consider the general application, which typically consists of multiple filters and inputs with a given depth.

Fig. 13.19: Original image (top); horizontal filter (middle); vertical filter (bottom).

Definition 13.6. *The lth layer of a neural network is said to constitute a* convolution layer *if it consists of a set of n_l filters $F^{l1}, \ldots, F^{ln_l}$ of size $\mathbb{R}^{f_l \times f_l \times d_l}$, such that it acts on a three-dimensional input array $A^{l-1} \in \mathbb{R}^{r_w^{l-1} \times r_h^{l-1} \times r_d^{l-1}}$, constrained by $d_l = r_d^{l-1}$, via the operation*

$$Z_{ijk}^l = \beta^{lk} + \left[F^{lk} \otimes A^{l-1} \right]_{ij}, \qquad (13.15)$$

where β^{lk} is an optional bias associated with the kth filter. Thus, each filter produces a distinct "slice" along the third dimension of the output Z^l.

The output of the layer is then given by the activated value

$$A^l = \phi(Z^l),$$

for activation function ϕ.

Typically, modern state-of-the-art leverages the ReLU as activation function, as it has been shown to be more efficient than sigmoidal activations, such as logistic and hyperbolic tangent, in terms of both speed and accuracy. Sometimes these two operations are viewed as separate, and one refers to a convolution layer and a ReLU layer, though these layers are typically glued together.

In addition to convolutional layers, CNNs also commonly contain *pooling layers*, the most common example of which is referred to as *max pooling*. Max pooling functions similar to the convolution operator, in which a grid of fixed size "scans" across the image. It differs from convolution, however, in that it outputs the maximum value from the image defined over the scope of the pooling grid. Thus, it serves as a moving maximum value across the image, surfacing a map of the most extreme pixel outputs within its visual field.

CNNs typically consist of many layers of convolution and pooling that alternate, followed by a sequence of fully connected layers. Several examples, including AlexNet and GoogLeNet architectures, are discussed in Gèron [2019], which also discusses implementation using TensorFlow and Keras.

13.4.3 Recurrent Neural Networks

Whereas convolutional neural networks are specialized to learn spatial patterns in multidimensional inputs, recurrent neural networks (RNNs) are specialized to learn patterns in temporal sequences, such as those that underlie applications with an intrinsic time element: time-series data, stock data, DNA sequences, and, most prominently, language applications (text completion, voice recognition, translation, etc.). In particular, RNNs break the independence assumption between successive inputs and outputs: *the* → __ (predict the next word) vs. *the cow jumped over the* → __ (predict the next word). Regardless of depth, no neural network can come close to predicting what comes after *the*; the best it could do is output a probability distribution over all nouns, with probability weighted relative to usage frequencies. But if the model can "remember" the immediate past—*the, cow, jumped, over*—the word that now follows *the* becomes most obvious: *moon*.

Recurrent neural networks are artificial neural networks the are comprised of a special type of neuron that passes its previous output into its next input.

Definition 13.7. *A recurrent neuron is an artificial neuron that reserves one component of its input interface to consume its previous output.*

A memory cell (or, simply, cell) is an artificial neuron with a hidden variable *that can change after each activation, is stored within the cell (i.e., as a form of short-term memory), and whose value is passed back into the cell with each successive input. In particular, if h represents the hidden variable, φ the neuron's activation function, and W_x and W_h two distinct sets of weights, we may write the output of a memory cell as*

$$y_t = \phi \left(W_x x_t + W_y h_{t-1} + \beta \right). \tag{13.16}$$

Here, β is the optional bias.

Though the concept of a recurrent neuron constitutes a special case of a memory cell, the distinction is a bit pedantic, and, in most cases, the two may be used interchangeably. The concept of a memory cell is depicted in Figure 13.20. We see that a recurrent neuron is just like an ordinary neuron, except that it has a loop from its output back into itself. Unraveling this over time, we obtain the illustration on the right, which shows how one activation's output connects to the input of the following input. The hidden variable h can be thought of as a form of short-term memory. When no prior output is available, a memory cell can except a `Null` value in its place.

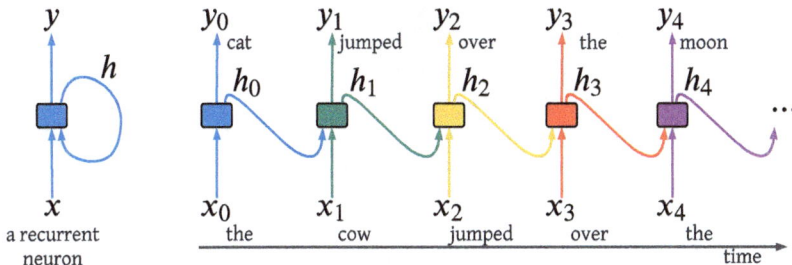

Fig. 13.20: A recurrent neuron (left) and its unraveled form (right).

Recurrent neural networks are then simply neural networks that are comprised of recurrent neurons or memory cells. Interestingly, the time element has implications for certain patterns of inputs and outputs. In particular, the inputs and outputs of an RNN might be either vectors or sequences, depending on the context of the problem. (Here, vector refers to the ordinary output of a vanilla neural network.) The illustration of Figure 13.20 constitutes a *sequence-to-sequence* network, such as predicting the next word in a sequence of text. In particular, each term of the input sequence has a corresponding output prediction, which ultimately constitutes an output sequence. Another example is an RNN that predicts the next day's stock price based on the current days' close: a sequence of stock prices is fed in, and a sequence of (one-day shifted) stock prices comes out.

A *sequence-to-vector network* ingests multiple inputs and hidden activations before a final (single) vector output is ultimately revealed. An example of this is *sentiment analysis*: the full sentence must be comprehended before the RNN can issue its final verdict: *happy*. An example of this is shown in the top half of Figure 13.21. Here, the grey outputs (y_0, y_1, y_2, and y_3) are ignored, and only the final output y_4 matters.

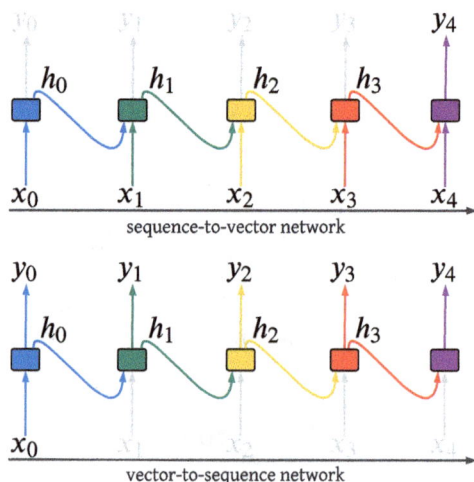

Fig. 13.21: A sequence-to-vector network (top) and a vector-to-sequence network (bottom). Grey variables and arrows are ignored.

Conversely, one might encounter a *vector-to-sequence networks*, which have a single input, followed by a number of activations that generate a sequence of outputs. For example, a single image array might be fed into the RNN as an input, and the RNN might activate multiples times to yield a text description of the image (*a cow jumping over the moon at night*). This is illustrated in the bottom half of Figure 13.21. Again, the grey inputs (x_1, x_2, x_3, and x_4) are ignored.

Finally, one might combine a sequence-to-vector network with a vector-to-sequence network to form a more elaborate structure called an *encoder–decoder*. This is common in language translation: the network needs to listen to the full sentence before it begins its translation, as the meaning of the sentence can be modified by the words at the end[4]. In this context, the sequence-to-vector network is referred to as the *encoder*, and the vector-to-sequence network is referred to as the *decoder*. Thus, a neural network can create an abstract neural-network encoding of the meaning of a sentence that transcends the language itself.

This is exactly what happened in 2016, when Google announced that that its new Neural Machine Translation (GNMT) system could translate between languages, even when it had never received training instances between those two specific languages (Schuster, *et al.* [2016]; for white paper, see Johnson, *et al.* [2016]). For example, GNMT learned how to translate between Japanese and Korean, without ever being shown any specific Japanese↔Korean pairs, as it already learned how to translate between Japanese↔English and between Korean↔English, using a single shared multilingual neural representation of the respective languages. At the time of the announcement, they were already translating between 10 different languages using a common abstract language representation.

RNNs are trained by unraveling the recurrent layers and backpropagating, a technique called *backpropagating through time*. Naturally, this is a delicate process. Due to the inherent instability in successively multiplying chains of the weight matrix, one easily encounters the same exploding and vanishing gradients issues prevalent in ordinary deep networks. In particular, gradients can sometimes vanish, which slows the learning system to a stop, or explode, which results in instability of values and ultimate divergence. To remedy this, we deploy the usual set of tricks, including good parameter initialization, dropout, etc. For details, see the usual references.

Long Short-Term Memory (LSTM) Cells

One drawback of the simple memory cells discussed thus far is their tendency to forget the past, making them impractical to handle patterns that extend beyond a handful of terms of a sequence. To remedy this, the concept of a *long short-term memory (LSTM) cell* was introduced by Hochreiter and Schmidhuber [1997]. Here, the cell's short-term memory, as represented by its hidden variable h, is augmented with a longer term memory, or *cell memory*, variable c. In addition, several devices, called *logic gates*, are used to control several basic operations, including which parts of the cell state should be updated or forgotten, and which parts should be output to the hidden state and output variable. Formally, we have the following.

---

[4] for example, in German it is common for verbs to occur at the end of the sentence: *I was out of popcorn, because I the movie yesterday with my friends at home watched had.*

Definition 13.8. *A* long short-term memory (LSTM) cell *is a memory cell that has an additional long-term hidden variable, called the* cell state *c, such that the basic operations of the cell consist of the calculation of three variables, f, i, and o, that control the* forget gate, input gate, *and* output gate, *respectively, and a variable g that depends on the previous hidden state and current input, via the relations*

$$f = \phi_\sigma \left(W_{xf} x_t + W_{hf} h_{t-1} + \beta_f \right) \tag{13.17}$$
$$i = \phi_\sigma \left(W_{xi} x_t + W_{hi} h_{t-1} + \beta_i \right) \tag{13.18}$$
$$o = \phi_\sigma \left(W_{xo} x_t + W_{ho} h_{t-1} + \beta_o \right) \tag{13.19}$$
$$g = \tanh \left(W_{xg} x_t + W_{hg} h_{t-1} + \beta_g \right), \tag{13.20}$$

where ϕ_σ is the sigmoid activation function, such that the cell state and hidden state are computed via

$$c_t = f \odot c_{t-1} + i \odot g \tag{13.21}$$
$$h_t = o \odot \tanh(c_t), \tag{13.22}$$

respectively. The variables f, i, o, and g are called helper variables *or* minions; *the first three are referred to as* gate controllers *and the fourth as the* penultimate output.

First, note that the intermediary variables f, i, o, and g are themselves identical to the output of a single memory cell (Equation (13.16)) with unique sets of input weights, hidden-state weights, and biases. When the input vector x and hidden state h are vectors, these constitute fully connected parallel layers, with a sigmoid activation function for the first three and a hyperbolic tangent activation for the fourth. The sigmoid activation function, which returns values between zero and one, is used for the first three, as they control which components of the cell state and hidden state should be modified. The tanh activation function is used for the fourth local variable g, which might be thought of as a sort of penultimate output; i.e., the computation of g alone could constitute the hidden state of an ordinary memory cell. These operations are illustrated graphically in Figure 13.22[5] (following the design from Gèron [2019]).

Now let's look at these operations in detail.

- $f \odot c_{t-1}$: since f is a vector of values on $(0, 1)$, this componentwise multiplication determines the degree to which each component of the cell state should be decayed, or "forgotten."
- $i \odot g$: since i is a vector of values on $(0, 1)$, this controls to what extend each component of the penultimate output g should be relevant in updating the cell state.

---

[5] Forgive the mild inconsistency: in the diagram, componentwise multiplication is symbolized by $\otimes$, whereas in the text we use $\odot$, as we have previously used the former symbol elsewhere to denote the outer product.

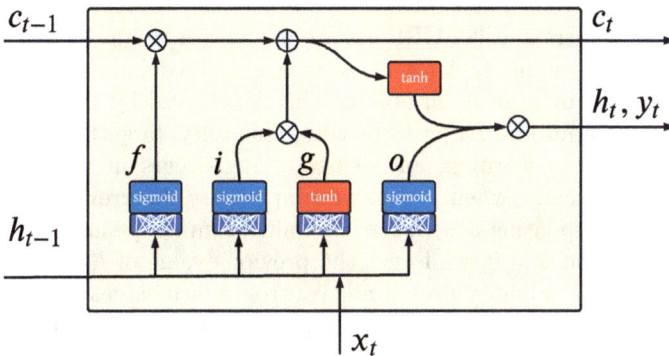

Fig. 13.22: Illustration of an LSTM cell. $\otimes$ and $\oplus$ denote componentwise multiplication and addition, respectively. The purple layers with the cross-hatch patterns sitting below f, i, g, and o are fully connected layers, each with their own set of weights.

- $c_t = f \odot c_{t-1} + i \odot g$: combines the previous two operations of forgetting a little bit of history and inputing new information, yielding the new cell state c_t.
- $h_t = o \tanh(c_t)$: since o is a vector of values on $(0, 1)$, this controls to what extend the activated cell state should be captured as hidden state, which equals the cell output $y_t = h_t$.

Layers of LSTM cells can now easily be constructed with minor modification to form a deep RNN. In particular, we would add a superscript l to weights and biases, hidden states and cell states, but replace the input vector x_t with h_t^{l-1}, which represents the output from the preceding layer, with the caveat that we define $h_t^0 = x_t$ to coincide with the actual input. In this way, the output of one layer feeds into the next.

Despite the success of LSTM cells in modern temporal machine-learning applications, they still struggle to remember sequences longer than $O(10^2)$, an improvement on simple memory cells by about a factor of ten. One approach to learn based on longer sequences is to apply a one-dimensional convolutional layer first, to generate a shorter temporal sequence that is better suited for LSTM-based RNNs.

GRU

A simplified form of LSTM has also been proposed, called the *gated recurrent unit* (GRU). The two models tend to have similar performance, with GRU having the advantage of being simpler to implement and more efficient. GRUs are also easier to train with smaller data sets; though using the tried-and-true LSTM is still the default choice for large data sets and longer sequences.

There are several main differences between GRUs and LSTMs. First, like vanilla recurrent cells, GRU cells only have a single hidden state h. In addition, the input and forget gates are merged into a single *update gate*, controlled by a local variable z. This is achieved by constraining the operations of input and forget to be complimentary: forgetting occurs only in the presence of learning, and remembering occurs in the presence of ignoring. Specifically, when $z = 1$, we simply pass the previous value of h along, without updating it with the new value from the penultimate state g. Conversely, when $z = 0$, we forget the previous value of h in favor of the new value from g. Finally, a *reset gate* controls which values of the previous hidden state are used in calculating the penultimate output g. Formally, we have the following.

Definition 13.9. *A gated recurrent unit (GRU) cell is a memory cell with three minions, consisting of two gate controllers, z and r, that control the* update gate *and* reset gate, *respectively, and a third minion g that depends on the previous hidden state, current input, and the reset gate, via the relations*

$$z = \phi_\sigma \left(W_{xz} x_t + W_{hz} h_{t-1} + \beta_z \right) \tag{13.23}$$

$$r = \phi_\sigma \left(W_{xr} x_t + W_{hr} h_{t-1} + \beta_r \right) \tag{13.24}$$

$$g = \tanh \left(W_{xg} x_t + W_{hg} \left(r \odot h_{t-1} \right) + \beta_g \right), \tag{13.25}$$

where ϕ_σ is the sigmoid activation function, such that the hidden state is computed via

$$h_t = z \odot h_{t-1} + (1 - z) \odot g, \tag{13.26}$$

From Equations (13.23)–(13.26), the workings of GRU are clear. The reset gate r controls the extent to which the previous value of h_{t-1} should be included in the calculation of the penultimate output g, as seen in Equation (13.25). The update gate z functions in the way we previously described, as shown by Equation (13.26). Thus, as was the case with LSTMs, the minions assist in a GRU cell's ability to carry out its basic operations.

Obviously, GRUs have smaller computational footprints than LSTMs and are therefore more efficient and easier to train. This comes at the cost, however, of losing a dedicated long-term memory variable. For this reason, LSTMs remain the prominent, but are still inefficient at learning sequences longer than around 100 terms. As mentioned previously, this can be remedied by combining LSTMs with one-dimensional convolutional layers to process extremely long temporal sequences. In particular, van den Oord, *et al.* [2016] introduced an architecture known as *WaveNet*, which actually stacks multiple convolutional layers, each one doubling the temporal spread of the previous layer, so that lower layers learn short-term patterns, whereas higher layers learn longer-term patterns.

Problems

13.1. Build a vector perceptron with softmax layer and train it on the iris dataset.

13.2. Code the `Activation` (Code Block 13.1) subclasses for the other five classic activation functions shown in Figure 13.6. *Bonus:* code the *leaky ReLU* activation function.

13.3. Explain why it is problematic to create a deep neural network using only linear activations.

13.4. Compute the output vector $\hat{p}$ for a three-class classification neural network for the 4-dimensional input vector $\langle 1, 2, 3, 4 \rangle$ and two layers with connection weights

$$W^1 = \begin{bmatrix} 1 & 0 & 0 & 0 & 0 \\ 1 & 1 & -1 & 1 & -1 \\ 1 & -1 & 1 & -1 & 1 \end{bmatrix} \quad \text{and} \quad \begin{bmatrix} 1 & 0 & 0 \\ 1 & -1 & 1 \\ -1 & 1 & 1 \\ 2 & -1 & 1 \end{bmatrix},$$

such that the first layer uses a ReLU activation and the second (output) layer uses a softmax transformation. (With the appropriate matrix and function definitions, this can be computed in Python using the single computation `softmax(W2@relu(W1@x))`, for appropriately defined functions `softmax` and `relu`.)

13.5. Draw the "non-exploded" version of the exploded neural network depicted in Figure 13.12.

13.6. Show that the derivatives δ_i^m indeed simplify to Equation (13.8) when carrying out the summation of the previous equation when computing the gradient over a softmax layer.

13.7. Suppose the true label of the training instance in Exercise 13.4 was $y = 1$ (i.e., $y = \langle 1, 0, 0 \rangle$). Use backpropagation to compute $\nabla_W L$ using the cross-entropy loss function.

13.8. Write an `AbstractLayer` class that defines the API common to the various layer implementations (i.e., `NeuronLayer`, `SoftmaxNeuronLayer`, `MatrixLayer`, `SoftmaxMatrixLayer`).

13.9. [**Project**] Update the `Loss`, `MatrixLayer` and `Activation` classes to accommodate batch gradient descent.

14

Rise of the Machines

In this chapter, we focus on a third branch of machine learning, separate from supervised and unsupervised learning, known as *reinforcement learning*. In this branch, the machine is given autonomy and allowed to make decisions and take actions. The machine then learns based on its ability to explore (try new things) and the environment's response to its choices. A classic favorite reference on reinforcment learning is Sutton and Barto [2018]. An additional reference is Kochenderfer [2015]. For information on the field of deep reinforcement learning, see Graesser and Keng [2019].

14.1 Reinforcement Learning

In this section, we will lay out the main mathematical formalism that underscores reinforcement learning. We will define the problem as a game between an agent and the environment, and define the solution in terms of policies and value functions. Finally, we introduce a set of optimality conditions.

14.1.1 Finite Markov Decision Processes

The goal of reinforcement learning is to learn how an agent can behave relative to a given environment, that is allowed to respond to the agent's action and give out rewards based on the agent's choices. We may formally regard an agent and its environment as follows.

Definition 14.1. *An* agent *is that which possesses agency.* Agency *is a duty to act.*

Actions inevitably occur over the passage of time, even if the agent doesn't *do* anything. Inaction is also an action. Absention from action is still a choice.

Definition 14.2. *An* environment *is an abstraction of an agent's world. An environment may have various* states *that may change in response to the action of the agent. The environment may at times give numerical* rewards *to the agent.*

The three basic quantities that describe the interaction between an agent and its environment are state, action, and reward.

Definition 14.3. *The* state space $\mathcal{S}$ *is the set of all possible states of the environment.*

The action space $\mathcal{A}$ *is the set of all permissible actions an agent might take. The* action bundle $\mathcal{A}_s \subset \mathcal{A}$ *is the subset of permissible actions available for each state $s \in \mathcal{S}$.*

The reward space $\mathcal{R}$ *is the set of possible rewards an agent might receive from the environment.*

Now that we have defined the state, action, and reward spaces, we may formally introduce the process that underlies much of modern-day reinforcement learning.

Definition 14.4. *A* Markov decision process (MDP) *is a quadruple $(\mathcal{S}, \mathcal{A}, \varphi, \varrho)$ consisting of a state space $\mathcal{S}$, an action space $\mathcal{A}$, a state-transition mapping $\varphi : \mathcal{S} \times \mathcal{A} \rightarrow \mathcal{P}(\mathcal{S})$, and a reward mapping $\varrho : \mathcal{S} \times \mathcal{S} \times \mathcal{A} \rightarrow \mathcal{P}(\mathbb{R})$ that defines a stochastic control process $\{(S_t, A_t, R_t)\}_{t \in \mathbb{N}}$, such that $\{A_t\}$ is a sequence of actions (i.e., controls) selected by an agent, $\{S_t\}$ is a sequence of states of the environment, which are governed by the* state-transition matrix (STM)

$$\varphi_s^a(s') = \mathbb{P}(S_{t+1} = s' | S_t = s, A_t = a), \tag{14.1}$$

and $\{R_t\}$ is a sequence of rewards, distributed according to

$$\varrho_{ss'}^a(r) = \mathbb{P}(R_{t+1} = r | S_t = s, A_t = a, S_{t+1} = s').$$

If the spaces $\mathcal{S}$ and $\mathcal{A}$ are finite, we refer to the MDP as finite.

Alternatively, if the full probability distribution over the reward space is not required, we may instead define our MDP by replacing ϱ with the expected-reward matrix (ERM) Υ, *which is the mapping $\Upsilon : \mathcal{S} \times \mathcal{S} \times \mathcal{A} \rightarrow \mathbb{R}$ defined by the expectations*

$$\Upsilon_{ss'}^a = \mathbb{E}[\varrho_{ss'}^a] = \mathbb{E}\left[R_{t+1} | S_t = s, A_t = a, S_{t+1} = s'\right]; \tag{14.2}$$

i.e., it is the expected reward at time $t + 1$, given the current state $S_t = s$ and action $A_t = a$ and the subsequent transition state $S_{t+1} = s'$.

Such a process is a *decision* process, as the sequence of actions are interpreted as choices to be made by an agent. The environment responds to those actions by periodically changing its state and issuing rewards (or

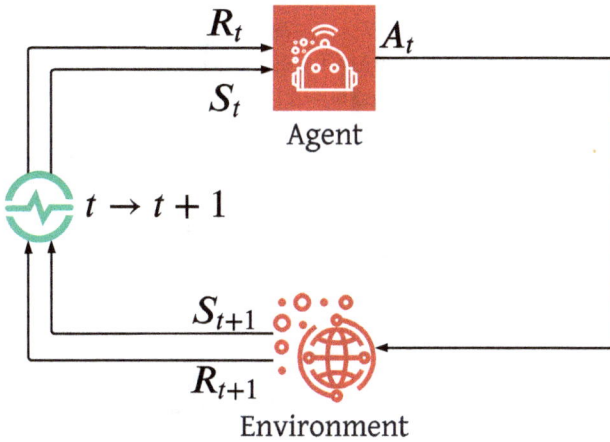

Fig. 14.1: Schematic of an agent and its environment.

penalties, if the reward is negative). This process is characterized in Figure 14.1.

So far, we have not placed any restrictions on the agent's actions; they can be totally uncontrolled and chaotic. The point, however, of reinforcement learning is to learn an optimal strategy for the agent—a method of behaving that is likely to maximize the long-term rewards paid out by the agent's environment. To quantify this, we need a way to mathematically describe how an agent might respond.

Definition 14.5. *Given a finite Markov decision process, a* policy *is a mapping* $\pi : \mathcal{S} \to \mathcal{P}(\mathcal{A})$, *such that the agent chooses an action relative to the probability distribution*

$$\pi_s(a) = \mathbb{P}(A_t = a | S_t = s). \tag{14.3}$$

The goal of reinforcement learning, as we shall see, is to determine an optimum policy, relative to a given objective function.

Given a state-transition matrix and an expected-reward matrix, we can use the former to marginalize over the transition states to obtain

$$\Upsilon_s^a = \mathbb{E}\left[R_{t+1} | S_t = s, A_t = a\right] = \sum_{s' \in \mathcal{S}} \varphi_s^a(s') \Upsilon_{ss'}^a. \tag{14.4}$$

Notice that we regard superscripts as actions and subscripts as states; when two subscripts are given, the second refers to the transition state. Moreover, we explicitly call out the function dependence of $\varphi_s^a(s')$ on s', since, for a given state $S_t = s$ and action $A_t = a$, this quantity represents a probability distribution over the state space corresponding to the possible values of the transition state $S_{t+1} = s'$.

Finally, we note that the state-transition matrix can be represented in matrix-form as $\Phi^a_{ss'} = \varphi^a_s(s')$, as shown in Figure 14.2. Here, $\Phi : \mathcal{A} \times \mathcal{S} \times \mathcal{S} \to [0, 1]$ is an alternate view that captures state-transition probabilities in tabular (matrix) form. This illustration is useful to keep in mind when coding a set of known transition probabilities. The highlighted parallelepiped represents all values corresponding to a single cell on the horizontal (a, s) plane. The probability constraint ensures these values sum to one.

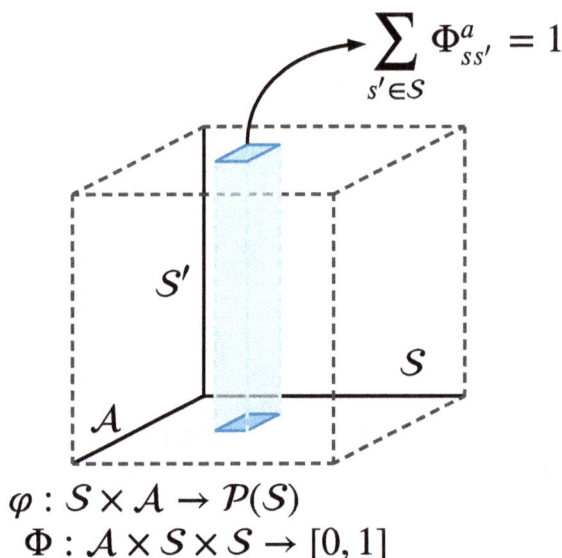

$$\sum_{s' \in S} \Phi^a_{ss'} = 1$$

$$\varphi : \mathcal{S} \times \mathcal{A} \to \mathcal{P}(\mathcal{S})$$
$$\Phi : \mathcal{A} \times \mathcal{S} \times \mathcal{S} \to [0, 1]$$

Fig. 14.2: Schematic of the state-transition matrix.

Episodic and Continuing Tasks

There are two main flavors of Markov decision processes: episodic and continuing.

Definition 14.6. An episodic task *is a reinforcement learning problem in which one seeks to learn based on a sequence of related MDPs, each consisting of only a finite number of steps. A continuing task is a reinforcement learning in which the underlying MDP is infinite in temporal scope (i.e., the stochastic processes go on indefinitely).*

For example, learning how to play chess would be considered an episodic task. Each individual chess game would constitute a separate *episode*. Based on the experience of playing many games over time, one seeks to become a better chess player. A continuing task, on the other hand, is simply a task that goes on indefinitely.

In order to unify notation, episodic tasks are typically endowed with a certain terminal state that only transitions into itself and never pays out any reward.

Definition 14.7. *An* absorbing state $s_0 \in S$ *is any state with the property that*

$$\varphi^a_{s_0}(s) = \mathbb{I}[s = s_0],$$

for all $a \in A$, and such that $R_{t+1} = 0$ whenever $S_t = s_0$.

In other words, an absorbing state only allows transitions into itself (hence the name) and never pays out any subsequent reward. In the context of chess, the absorbing state may be considered a fictitious state following a *checkmate*, that state being the status: *the game is over*.

By including an absorbing state with any episodic task, we may continue to use the same notation (e.g., infinite sums over future rewards) as we use for continuing tasks.

14.1.2 Value Functions

Much of reinforcement learning reduces to the task of learning certain *value functions*, which express the *value* of a given state (or state–action pair) relative to a given policy.

Present Value of Future Returns

Typically, we wish to maximize the *total expected return*, which consists of summing over all future rewards. For continuing tasks, however, this sum tends to diverge, since an infinite time horizon can easily give rise to an infinite return. This is resolved by *discounting* future rewards with a given *discount rate* γ, to account for the *time value of money*. Formally, we have the following.

Definition 14.8. *The* return *of a Markov decision process at time t is the discounted sum of future rewards, as given by*

$$G_t = R_{t+1} + \gamma R_{t+2} + \gamma^2 R_{t+3} + \cdots = \sum_{k=0}^{\infty} \gamma^k R_{t+k+1}. \tag{14.5}$$

where the parameter $\gamma \in [0, 1]$ is the discount rate.

Naturally, as long as the reward sequence is bounded, the return will be finite for $\gamma < 1$. If $\gamma = 0$, the agent is only concerned with maximizing its immediately subsequent reward. This is, in general, not optimal, as acting to optimize an immediate reward can limit the collection of future rewards; see, for example, the Stanford marshmallow experiment (Mischel and Ebbesen [1970]).

In episodic tasks, it is possible for a reward to be zero until the penultimate state; i.e., penultimate if we view the final state as a zero-reward absorbing state. For example, in the game of chess, the reward can be zero or one, depending if the agent win's or loses. This illustrates the absolute necessity in the ability to backpropagate future rewards to much earlier behavior.

State and State–Action Value Functions

Value functions are ubiquitous in reinforcement learning. They are functions that prescribe a value for an agent to be in a given state, or to take a given action from a given state. Naturally, this notion of value relies on how the agent intends to act for the duration of the game, as quantified by its policy. We therefore reference the two main types of value functions in relation to a given policy.

Definition 14.9. *The* state value function *(or, simply,* value function*) of a state $s \in \mathcal{S}$ under a policy π is the function*

$$v_\pi(s) = \mathbb{E}_\pi\left[G_t | S_t = s\right] = \mathbb{E}_\pi\left[\sum_{k=0}^{\infty} \gamma^k R_{t+k+1} \,\middle|\, S_t = s\right], \qquad (14.6)$$

where the expectation is taken relative to the policy π; i.e., it represents the expected value assuming that the agent acts according to the policy π for all future time.

The state-value function therefore represents the value (i.e., expected return) for the agent to be in a given state, assuming the agent follows the policy π for all time thereafter. It is also useful to quantify the value of an agent to be in a given state *and* make a given action, and then follow the policy for all time thereafter. We define this as follows.

Definition 14.10. *The* state-action value function *of a state $s \in \mathcal{S}$ and action $a \in \mathcal{A}$ under a policy π is the expected return*

$$q_\pi(s, a) = \mathbb{E}_\pi\left[G_t | S_t = s, A_t = a\right], \qquad (14.7)$$

where G_t is the expected return given by Equation (14.5).

This is a slight twist, as it allows consideration of the agent to take any action from state s, making explicit reference to the action taken. Since the state value function assumes A_t is taken relative to the given policy, i.e., $A_t \sim \pi_s$, it is related to the state-action value function by the equation

$$v_\pi(s) = \sum_{a \in \mathcal{A}} \pi_s(a) q_\pi(s, a). \tag{14.8}$$

However, the state-action value function is agnostic as to how the *present action* A_t is chosen, as long as all *future* actions are taken according to the policy in question.

The Bellman equation

The Bellman equation is simply a recurrence relation that expresses the relationship between the value of a state and the value of its successor state. The unique solution to this equation is the state value function. We begin with a lemma.

Lemma 14.1. *The state-action value function of an MDP under a policy π and discount rate γ is related to the state value function and transition probabilities by the relation*

$$q_\pi(s, a) = \sum_{s' \in \mathcal{S}} \varphi_s^a(s') \left[\Upsilon_{ss'}^a + \gamma v_\pi(s') \right]. \tag{14.9}$$

Proof. We begin by noting

$$q_\pi(s, a) = \mathbb{E}_\pi[G_t | S_t = s, A_t = a] = \mathbb{E}_\pi[R_{t+1} + \gamma G_{t+1} | S_t = s, A_t = a],$$

which follows easily enough from Equation (14.5).

Now, the expected value of the next reward can be obtained by applying the law of total expectation by considering all possible transition states, from which we obtain

$$\mathbb{E}_\pi[R_{t+1} | S_t = s, A_t = a] = \sum_{s' \in \mathcal{S}} \varphi_s^a(s') \Upsilon_{ss'}^a,$$

which is equivalent to Equation (14.4).

Next, we may express the quantity $\mathbb{E}_\pi[G_{t+1} | S_t = s, A_t]$ by again considering all possible transition states, thereby obtaining

$$\mathbb{E}_\pi[G_{t+1} | S_t = s, A_t = a] = \sum_{s' \in \mathcal{S}} \varphi_s^a(s') \mathbb{E}_\pi[G_{t+1} | S_{t+1} = s'] = \sum_{s' \in \mathcal{S}} \varphi_s^a(s') v_\pi(s').$$

The result follows. □

There are two fundamentals ways in which we might apply Equation (14.8) to Equation (14.9). The first is to replace $q_\pi(s,a)$ in Equation (14.8) with the right-hand side of Equation (14.9). This result is known as the *Bellman equation*, as given by the following theorem. But note the second way: we could also replace $v_\pi(s')$ in the summand of Equation (14.9) with the right-hand side of Equation (14.8), evaluated, of course, at s'. This yields a recurrence relation for the state-action value function.

Proposition 14.1 (Bellman equation). *The state value function of an MDP under a policy π and discount rate γ must satisfy the following recurrence relation*

$$v_\pi(s) = \sum_{a \in \mathcal{A}} \sum_{s' \in \mathcal{S}} \pi_s(a) \varphi_s^a(s') \left[\Upsilon_{ss'}^a + \gamma v_\pi(s') \right]. \qquad (14.10)$$

Proof. This follows immediately by applying the relation given by Equation (14.8) to Equation (14.9).   □

Backup diagrams are useful in visualizing how Bellman's equation propagates information from the future *back* to a given state, as shown in Figure 14.3. By convention, empty circles represent states and filled circles represent actions. This diagram shows a given state $S_t = s$, and all possible actions and transition states. The Bellman equation represents an average across all possible outcomes, using the policy to weight the probability of given actions and the state-transition matrix to weight the probability of a given transition state, given the current state and action.

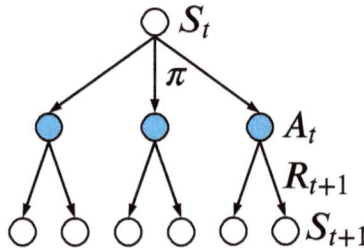

Fig. 14.3: Backup diagram for the Bellman equation.

14.1.3 Optimality

In this section, we introduce the notion of an optimal policies and optimal value functions. We then discuss the Bellman optimality equations.

Optimal Policies and Optimal Value Functions

State value functions define a partial ordering over policies, since we can define $\pi \geq \pi'$ if and only if $v_\pi(s) \geq v_{\pi'}(s)$ for all $s \in \mathcal{S}$. This naturally gives rise to the notion of an optimal policy.

Definition 14.11. *Given a finite MDP, an* optimal policy π_* *is a policy that satisfies* $\pi_* \geq \pi$, *for all other policies* π; *in other words,* $v_{\pi_*}(s) \geq v_\pi(s)$ *for all* $s \in \mathcal{S}$ *and all policies* π.

The optimal state value function, *denoted* v_*, *is defined by the relation*

$$v_*(s) = \max_\pi v_\pi(s). \tag{14.11}$$

Similarly, the optimal state-action value function, *denoted* q_*, *is defined by the relation*

$$q_*(s, a) = \max_\pi q_\pi(s, a). \tag{14.12}$$

The optimal state value and state-action value functions are the value functions of an optimal policy.

Bellman Optimality Equations

Since v_* and q_* are value functions for the optimal policy, they too must satisfy a similar set of self-consistency equations provided by the Bellman's equation. Since these value functions are, however, optimal, we may express their respective recurrence relations without regard to the specific policy. The result is known as the Bellman optimality equations, and is given as follows.

Proposition 14.2 (Bellman optimality equations). *Given a finite MDP, the optimal state value function satisfies the recurrence relation*

$$v_*(s) = \max_{a \in \mathcal{A}_s} \sum_{s' \in \mathcal{S}} \varphi_s^a(s') \left[\Upsilon_{ss'}^a + \gamma v_*(s') \right]. \tag{14.13}$$

Similarly, the state-action value function satisfies the recurrence relation

$$q_*(s, a) = \sum_{s' \in \mathcal{S}} \varphi_s^a(s') \left[\Upsilon_{ss'}^a + \gamma \max_{a' \in \mathcal{A}_{s'}} q_*(s', a') \right]. \tag{14.14}$$

Proof. Equation (14.13) follows from Equation (14.9), when expressed for an optimal policy, coupled with the observation that

$$v_*(s) = \max_{a \in \mathcal{A}_s} q_*(s, a);$$

i.e., the optimal value of a state s must be the maximum value, taken over all possible actions, of the optimal state-action value function anchored to the same state.

Equation (14.14) is also obtained from Equation (14.9), again rewritten for an optimal policy, except this time we use the preceding equation to replace $v_*(s')$ in the summand. $\qquad \square$

Optimal Policies from Optimal Value Functions

Once we know the optimal value function v_*, we can determine an optimal policy π_* as follows. For each state $s \in \mathcal{S}$, there will be at least one action that yields the maximum value in the Bellman optimality Equation (14.13). Any policy that assigns a zero probability to actions outside of the set of actions that achieve this maximum therefore constitutes an optimal policy. We may state this result formally as follows.

Proposition 14.3. *Let v_* be the optimal state value function for a finite MDP. For each state $s \in \mathcal{S}$, let*

$$\mathcal{A}_s^* = \arg\max_{a \in \mathcal{A}_s} \sum_{s' \in \mathcal{S}} \varphi_s^a(s') \left[\Upsilon_{ss'}^a + \gamma v_*(s') \right] \tag{14.15}$$

represent the set of actions that maximize the expected return via the Bellman optimality condition. Any policy that satisfies the condition

$$\pi_s = \mathrm{Unif}(\mathcal{A}_s^*) = \frac{1}{|\mathcal{A}_s^*|} \mathbb{I}[a \in \mathcal{A}_s^*], \tag{14.16}$$

for all $s \in \mathcal{S}$, is an optimal policy.

In other words, optimal policies are *greedy* with respect to the optimal value function (Definition 4.20). This is because the total expected future reward has been *pulled back* into a value function at each state.

14.1.4 Examples

Gridworld

Example 14.1. A classic example is that of the so-called *gridworld* problem. A *gridworld* is a two-dimensional state space arranged as a grid with a particular geometry. For example, a 5×5 gridworld is shown in Figure 14.4. (Gridworlds are typically, but not necessarily, rectangular.) The rules of gridworld are simple: each state has four possible actions corresponding to the four compass directions. The transition state is 100% the direction specified by the action, with the exception that the transition state is unchanged if either the action points outside of the gridworld or the state is a prescribed absorbing state. The gridworld shown in Figure 14.4 has a single absorbing state at $s_0 = (2, 2)$.

Furthermore, we specify a deterministic reward of -1 for each transition, with the exception, of course, that transitions from the absorbing state to itself have zero value. The optimal policy is shown by the arrows: since each move has a cost, it is optimal to move to the absorbing state as quickly as possible. The optimal state values, using a discount rate of $\gamma = 0.5$, are also given in each square. ▷

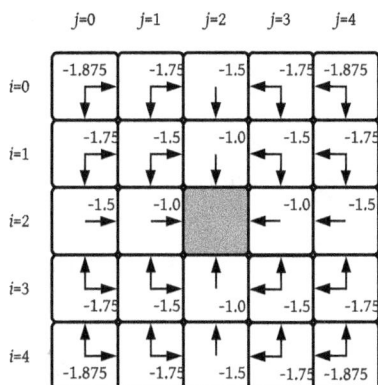

Fig. 14.4: Optimal policy and value function (using discount $\gamma = 0.5$) for the gridworld problem of Example 14.1.

Paper–Scissors–Rock

The following is an example of *adversarial reinforcement learning*, in which two agents compete for a reward. This describes most games: tic-tac-toe, checkers, chess, etc. We may still leverage the reinforcement-learning framework developed thus far by focusing on one agent, and viewing the actions and consequences of the adversarial agent as part of the environmental response.

Example 14.2. In the game of *paper–scissors–rock*, two opponents simultaneously reveal their selection (from the three namesakes) and a winner is determined according to the following binary relation: scissors cut paper, rock smashes scissors, paper covers rock. We may encode these choices using ternary ($\mathbb{Z}_3 = \{0, 1, 2\}$) by assigning a values (0 to paper, 1 to scissors, and 2 to rock) and defining the following binary relation:

$$0 \prec 1, \qquad 1 \prec 2, \qquad \text{and} \qquad 2 \prec 0.$$

Naturally, addition is taken to be mod 3, such that $2 + 1 = 0$.

Using the ternary encoding, we may represent our state space as the set $\mathcal{S} = \mathbb{Z}_3^2$ of two-dimensional ternary vectors, which may be represented as a length-two ternary string. The first component corresponds to the agent's choice and the second the adversary's. For example, the state $s = 21$ represents the agent has chosen *rock* and the adversary has chosen *scissors*. The action space is simply $\mathcal{A} = \mathbb{Z}_3$. Finally, we may specify a deterministic reward R for a transition state $S' = ij$ as $R = -1$ whenever $i \prec j$, $R = +1$ whenever $i \succ j$, and $R = 0$ whenever $i = j$.

In order to specify the transition dynamics, let us make a simple model about our adversary: when they lose or tie, they select their next action with equal probability, but when they win, they choose to remain at their current state with a 50% probability, otherwise they transition to one of the other two states with equal probability. The transition state is determined by joining the agent's action and the adversary's action into a single length-two ternary string. These probabilities are tabulated in Table 14.1.

$R_t = r$	$S_t = ij$	$A_t = a$	$\varphi_{ij}^a(a0)$	$\varphi_{ij}^a(a1)$	$\varphi_{ij}^a(a2)$
0	00	a	1/3	1/3	1/3
−1	01	a	1/4	1/2	1/4
+1	02	a	1/3	1/3	1/3
+1	10	a	1/3	1/3	1/3
0	11	a	1/3	1/3	1/3
−1	12	a	1/4	1/4	1/2
−1	20	a	1/2	1/4	1/4
+1	21	a	1/3	1/3	1/3
0	22	a	1/3	1/3	1/3

Table 14.1: State transition probabilities for the paper–scissors–rock example; $\varphi_{ij}^a(bj') = 0$ whenever $a \neq b$.

Moreover, since the reward depends deterministically on the transition state, which, itself, may be expressed in terms of the action, we may write

$$\Upsilon_{(ij)(aa)}^a = 0, \qquad \Upsilon_{(ij)(a(a+1))}^a = -1, \qquad \Upsilon_{(ij)(a(a-1))}^a = +1.$$

Notice we can leverage the asymmetry present when our agent loses (and the adversary wins) to determine an optimal policy. For this case, the adversary is more likely to remain in its current state, so that our optimal policy is to select an action that would defeat our adversary in its current state; i.e.,

$$\pi_{ii}^*(a) = \pi_{i(i-1)}^*(a) = 1/3 \qquad \text{and} \qquad \pi_{i(i+1)}^*(i - 1) = 1,$$

for $i \in \mathbb{Z}_3$. In other words: our agent should take random actions whenever it wins or ties, and transition to the missing state if it loses.   ▷

14.2 Dynamic Programming

Dynamic programming is a technique that can be used to determine optimal policies. We have already discussed one instance of dynamic programming in the backpropagation algorithm for neural networks. Like its deep-learning counterpart, dynamic programming for reinforcement learning consists of two steps: an evaluation (prediction) step and an update (improvement) step. Dynamic programming algorithms then seek to iteratively improve our estimation for an optimal policy or the optimal value function by alternating between these two steps over time. This approach is, however, largely of theoretical interest, as it relies on perfect knowledge of the transition probabilities and reward function. It is also computationally expensive. Nevertheless, it serves as a foundational model for the other, more practical, algorithms we will discuss through the remainder of the chapter.

14.2.1 Policy Iteration

Policy iteration functions similar to both backpropagation (Algorithm 13.4) and the Kalman filter (Bayes–Kalman filter discussed in Algorithm 10.1): it is a two-step process that involves a *prediction step* and an *update step*. In policy iteration, one alternates between a prediction step, which seeks to solve for the value function of a given policy, and an update step, which seeks to update the policy to bring it closer to the optimal policy. These iterations continue until the sequence of policies converges on an optimal policy.

This basic policy-iteration algorithm is shown in Algorithm 14.1. We begin by initializing an arbitrary value function and policy, except that the value function should have a zero value on any absorbing states, if applicable. We then iterate between the two operations `evalPolicy` and `updatePolicy`, which we will describe momentarily. We note, however, that the `evalPolicy` method is itself iterative, and that its second argument, which represents the starting "guess" for the value function, is optional. In practice, it is much faster to pass in the previous value function (as shown in the algorithm), as the `evalPolicy` method will tend to require significantly

---

Algorithm 14.1: Policy iteration.

Input: An MDP $(\mathcal{S}, \mathcal{A}, \varphi, \Upsilon)$
Output: An optimal policy π_* and optimal value function v_*
1 Initialize an arbitrary function $V : \mathcal{S} \to \mathbb{R}$
2 Initialize an arbitrary policy $\pi : \mathcal{S} \to \mathcal{P}(\mathcal{A})$
3 Set $V(s_0) = 0$ for any absorbing state s_0
4 Set *converged* = **False**
5 **while not** *converged* **do**
6 $\quad$ Set $V = $ `evalPolicy`(π, V)
7 $\quad$ Set $\pi' = $ `updatePolicy`(V)
8 $\quad$ Set *converged* $= \mathbb{I}[\pi = \pi']$
9 $\quad$ Set $\pi = \pi'$
10 **end**
11 **return** (π, V)

---

fewer iterations to converge. The idea is that if the policy changes just a little bit, then so too will its corresponding value function.

This iteration process terminates once it fails to yield an improvement on the current policy.

Policy Evaluation (Prediction)

The objective of *policy evaluation* is to determine the state value function v_π associated with a given policy π and a given MDP. If we know the transition probabilities and reward matrix, Equation (14.10) represents a linear system of $|\mathcal{S}|$ equations in $|\mathcal{S}|$ unknowns, which can be solved using standard techniques from linear algebra. It is more common, however, in practice, to use an iterative approach, as outlined in Algorithm 14.2. Here, we simply use the right-hand side of Equation (14.10) as an update rule to iteratively improve our estimate for the value $V(s)$ of each state $s \in \mathcal{S}$.

Note that we are updating the values of our estimate $V(s)$ of the state value function as we loop through the state space. Other than the first state in this loop, this update is therefore a blend of old values and new values. This tends to converge faster than the alternative—only applying the updates once all the calculations have been made—as one is utilizing each updated value as soon as it becomes available.

Algorithm 14.2: evalPolicy(π, V) policy evaluation.

Input: An MDP $(\mathcal{S}, \mathcal{A}, \varphi, \varUpsilon)$;
   A given policy π;
   An initial guess (optional) $V : \mathcal{S} \to \mathbb{R}$;
   A discount factor $\gamma \in [0, 1]$;
   An error tolerance ϵ
Output: The value function v_π of the given policy

1 **if** V **is Null then**
2 | Initialize an arbitrary function $V : \mathcal{S} \to \mathbb{R}$
3 | Set $V(s_0) = 0$ for any absorbing state s_0
4 **end**
5 Set $\Delta = 2\epsilon$
6 **while** $\Delta > \epsilon$ **do**
7 | Set $\Delta = 0$
8 | **for** s **in** $\mathcal{S}$ **do**
9 | | Set $v = V(s)$
10 | | Set $V(s) = \sum_{a \in \mathcal{A}} \sum_{s' \in \mathcal{S}} \pi_s(a) \varphi_s^a(s') \left[\varUpsilon_{ss'}^a + \gamma V(s') \right]$
11 | | Set $\Delta = \max(\Delta, |v - V(s)|)$
12 | **end**
13 **end**
14 **return** V

Finally, we notice the optional input argument V, which constitutes an optional "initial guess" to the value function. As previously mentioned, this is extremely useful in the process of policy iteration, as the *prediction step* itself represents multiple sweeps through state space, which can be quite large. By initializing the algorithm with something close to the true value function, as opposed to a completely random initialization, we require fewer loops through state space to achieve convergence.

Policy Improvement (Update)

A state value function for a given policy can be used to determine a more optimal policy. Essentially, this is achieved by defining a new policy that selects the action that maximizes Equation (14.9) for each state. To arrive at this conclusion, we begin with the following.

Theorem 14.1 (Policy Improvement Theorem). *Given a Markov decision process $(\mathcal{S}, \mathcal{A}, \varphi, \Upsilon)$ and two policies π and π', such that*

$$v_\pi(s) \leq \sum_{a \in \mathcal{A}_s} \pi'_s(a) q_\pi(s, a), \tag{14.17}$$

for all $s \in \mathcal{S}$, then $\pi' \geq \pi$, in the sense that $v_{\pi'}(s) \geq v_\pi(s)$ for all $s \in \mathcal{S}$.

Essentially, Theorem 14.1 says that if the expected value of taking the first step according to the new policy π', and then following the original policy π thereafter, is greater than the value of strictly following the original policy π, then the new policy must be more optimal than the original.

Proof. We proceed by induction. First, let us define a sequence of policies $\pi_0, \pi_1, \pi_2, \ldots$, such that π_k is defined by following the new policy π' for the first k steps and then following the original policy π thereafter. Note that $\pi_0 = \pi$ and

$$\lim_{k \to \infty} \pi_k = \pi'.$$

For the induction step, we note that the right-hand side of Equation (14.17) is equivalent to

$$\sum_{a \in \mathcal{A}_s} \pi'_s(a) q_\pi(s, a) = v_{\pi_1}(s),$$

due to Equation (14.8). This implies that $\pi_0 \leq \pi_1$.

Next, consider the state value function for policy π_k, which may be expressed as

$$v_{\pi_k}(s) = \mathbb{E}_{\pi'} \left[R_1 + \gamma R_2 + \cdots + \gamma^{k-1} R_k + \gamma^k v_\pi(S_k) \middle| S_0 = s \right].$$

Now, from Equation (14.17), and any possible state $S_k = s_k$, we have

$$v_\pi(s_k) \leq \sum_{a \in \mathcal{A}_{s_k}} \pi'_{s_k}(a) q_\pi(s_k, a) = \mathbb{E}_{\pi'} \left[R_{k+1} + \gamma v_\pi(S_{k+1}) \middle| S_k = s_k \right].$$

However, this implies that

$$v_{\pi_k}(s) \leq \mathbb{E}_{\pi'} \left[R_1 + \cdots + \gamma^k R_{k+1} + \gamma^{k+1} v_\pi(S_{k+1}) \middle| S_0 = s \right] = v_{\pi_{k+1}}(s),$$

so that $\pi_k \leq \pi_{k+1}$.

We have therefore constructed a monotonic increasing sequence of policies that starts with the original π and converges to the new policy π'. The result that $\pi \leq \pi'$ follows.   □

In order to apply this, let us define the following.

Definition 14.12. *Given a Markov decision process $(\mathcal{S}, \mathcal{A}, \varphi, \Upsilon)$ and an arbitrary state-value function $q(s, a)$, we define the* optimal-action bundle *relative to q, denoted $\mathcal{A}_s^q$, as the mapping $\mathcal{A}^q : \mathcal{S} \to \mathcal{A}$ defined by*

$$\mathcal{A}_s^q = \arg\max_{a \in \mathcal{A}_s} q(s, a) \tag{14.18}$$

for each $s \in \mathcal{S}$. Note $\mathcal{A}_s^q \subset \mathcal{A}_s$.

Corollary 14.1. *Given a Markov decision process $(\mathcal{S}, \mathcal{A}, \varphi, \Upsilon)$, a policy π, and its corresponding state-action value functions q_π, then the new policy π', defined by*

$$\pi'_s = \mathrm{Unif}(\mathcal{A}_s^{q_\pi}), \tag{14.19}$$

for all $s \in \mathcal{S}$, where $\mathcal{A}_s^{q_\pi}$ is the optimal-action bundle relative to q_π, is more optimal than the original policy π; i.e., $\pi' \geq \pi$.

Note 14.1. For a system with known state-transition dynamics, the optimal-action bundle may be expressed via the state value function by using Equation (14.9) to express Equation (14.18) in the equivalent form

$$\mathcal{A}_s^{q_\pi} = \arg\max_{a \in \mathcal{A}_s} \sum_{s' \in \mathcal{S}} \varphi_s^a(s') \left[\Upsilon_{ss'}^a + \gamma v_\pi(s') \right], \tag{14.20}$$

for all $s \in \mathcal{S}$.   ▷

Proof. The argument of the arg max operator of Equation (14.20) is equivalent to the state-action value function $q_\pi(s, a)$, due to Equation (14.9). Therefore, the right-hand side of Equation (14.17) must equal the maximum

$$\sum_{a \in \mathcal{A}_s} \pi'_s(a) q_\pi(s, a) = \max_{a \in \mathcal{A}_s} q_\pi(s, a),$$

since the support of the new policy π' is over the actions that maximize the state-action values. The result follows by Theorem 14.1.   □

This result is the basis for the policy improvement algorithm, as given by Algorithm 14.3. Note that the result still holds if we replace Equation (14.19) by any probability distribution whose support is A_s^π; i.e., that selects only from optimal actions. We simply choose the uniform distribution for good measure[1].

14.2.2 Value Iteration

As we have seen, policy iteration involves alternating between a prediction step, used to evaluate the current policy, and an update state, used to improve upon the current policy. The first step is, itself, iterative, as it requires multiple sweeps through state space before it can converge on the correct value function (within some absolute error tolerance).

Value iteration seeks to simplify this process by combining the two steps, and only performing *a single sweep* through state space for the prediction step. The result is shown in Algorithm 14.4.

Note, in particular, that line 8 of this algorithm replaces the *two* separate computations in policy evaluation and policy improvement of computing the new value of the state value function and the new policy. The reason these two operations can be combined is that the policy is determined by computing the arg max, and then the value function is determined by evaluating the same summation for the new policy, which must therefore yield the simple maximum value. In this way, value iteration works on iteratively improving the state value function and does not require computation of the policy until the very end.

A further illumination can be had if we define the estimated state-action value function in a parallel form to Equation (14.9); i.e., if we define

$$Q(s, a) = \sum_{s' \in \mathcal{S}} \varphi_s^a(s') \left[\Upsilon_{ss'}^a + \gamma V(s') \right]. \tag{14.21}$$

Algorithm 14.3: updatePolicy(V) policy improvement.

Input: An MDP $(\mathcal{S}, \mathcal{A}, \varphi, \Upsilon)$;
        State value function V;
        A discount factor $\gamma \in [0, 1]$
Output: Improved policy π

1 **for** s in $\mathcal{S}$ **do**

2 $\quad$ Set $\pi_s = \mathrm{Unif}\left(\underset{a \in \mathcal{A}_s}{\arg\max} \sum_{s' \in \mathcal{S}} \varphi_s^a(s') \left[\Upsilon_{ss'}^a + \gamma V(s') \right] \right)$

3 **end**
4 **return** π

---

[1] lol.

The state-value update in the policy valuation Algorithm 14.2 essentially averages the state-action values relative to the given policy:

$$V(s) = \sum_{a \in \mathcal{A}} \pi_s(a) Q(s, a). \tag{14.22}$$

In this way, the policy π remains unchanged and we successively get closer to the true state value function we are trying to estimate. Value iteration, on the other hand, further selects the optimal action in this update, by taking

$$V(s) = \max_{a \in \mathcal{A}_s} Q(s, a), \tag{14.23}$$

which is policy agnostic. Thus, Equation (14.22) represents the policy evaluation of Algorithm 14.2, whereas Equation (14.23) represents the value iteration of Algorithm 14.4, as it updates both our estimate for V and the policy in one swoop.

14.2.3 Examples

Paper–Scissors–Rock

Example 14.3. Let us continue the paper–scissors–rock problem of Example 14.2. The code for value iteration is given in Code Block 14.1.

Algorithm 14.4: Value iteration.

Input: An MDP $(\mathcal{S}, \mathcal{A}, \varphi, \Upsilon)$;
      A discount factor $\gamma \in [0, 1]$;
      An error tolerance ϵ
Output: A near optimal policy $\pi \approx \pi_*$ and its associated value
      function v_π

1 Initialize an arbitrary function $V : \mathcal{S} \to \mathbb{R}$
2 Set $V(s_0) = 0$ for any absorbing state s_0
3 Set $\Delta = 2\epsilon$
4 **while** $\Delta > \epsilon$ **do**
5   Set $\Delta = 0$
6   **for** s in $\mathcal{S}$ **do**
7   Set $v = V(s)$
8   Set $V(s) = \max\limits_{a \in \mathcal{A}_s} \sum\limits_{s' \in \mathcal{S}} \varphi_s^a(s') \left[\Upsilon_{ss'}^a + \gamma V(s') \right]$
9   Set $\Delta = \max(\Delta, |v - V(s)|)$
10   **end**
11 **end**
12 Set $\pi_s = \text{Unif}\left(\arg\max\limits_{a \in \mathcal{A}_s} \sum\limits_{s' \in \mathcal{S}} \varphi_s^a(s') \left[\Upsilon_{ss'}^a + \gamma V(s') \right] \right)$
13 **return** (π, V)

```python
# Initialize States, Actions, Rewards, and State Transitions
actions = [str(i) for i in range(3)]
states = [i+j for i in actions for j in actions]
rewards = [int((int(s[1])+1)%3 == int(s[0])) - int((int(s[0])+1)%3
    == int(s[1])) for s in states]
phi = np.zeros((3,9,9))
for i, a in enumerate(actions):
    for j, s0 in enumerate(states):
        for k, s1 in enumerate(states):
            if a != s1[0]:
                continue
            if rewards[j] == -1:
                phi[i, j, k] = 0.25 + 0.25 * int(s0[1] == s1[1])
            else:
                phi[i, j, k] = 1/3
# Set parameters
epsilon, gamma = 0.001, 0.5
n_states, n_actions = 9, 3
delta = 2 * epsilon
V, dV, pis = np.zeros(9), None, []
# Value Iteration
while delta > epsilon:
    delta = 0
    pi = DataFrame(0, index=states, columns=range(3))
    for j in range(n_states):
        v = V[j]
        Q = np.zeros(3)
        for i in range(n_actions):
            for k in range(n_states):
                Q[i] += phi[i, j, k] * (rewards[k] + gamma * V[k])
        V[j] = max(Q)
        delta = max(delta, abs(v-V[j]))
        Q = Q.round(3)
        best_actions = np.argwhere(Q==max(Q)).flatten()
        for a in best_actions:
            pi.loc[states[j], a] = 1/len(best_actions)
    pis.append(pi)
    if dV is None:
        dV = Series(V*1, index=states)
    else:
        dV = pd.concat((dV, Series(V, index=states)), axis=1,
            ignore_index=True)
```

Code Block 14.1: Value iteration for paper–scissors–rock problem.

We begin by defining the state and action space, and the corresponding deterministic rewards, along with the state-transition matrix in lines 2–14. Recall that states consist of a length-two ternary string, whose first component corresponds to the action (paper, scissors, or rock) of the agent, and whose second component corresponds to the action of the adversary. Transition probabilities are therefore only defined when the action a coincides with the first part of the transition state s'. Moreover, when the agent loses, the adversary is 50% likely to choose its current state as its next action, as shown on line 12, but otherwise equally likely to choose any action (line 14).

Value iteration is then achieved through the loop on lines 21–40. Here, we leverage the update form of Equation (14.23). Since this is computed for each state iteratively, we only have to store the state-action values for the three actions in our array Q. We also calculate the policy within each step, just as means of illustration, though the policy is typically not calculated for value iteration until the very end.

state	0	1	2	3	$\pi_s^3(0)$	$\pi_s^3(1)$	$\pi_s^3(2)$
00	0.0	0.0724	0.079	0.0799	0.333	0.333	0.333
01	0.25	0.3124	0.319	0.3199	0	0	1
02	0.0417	0.0724	0.079	0.0799	0.333	0.333	0.333
10	0.0486	0.0762	0.0795	0.0799	0.333	0.333	0.333
11	0.0486	0.0762	0.0795	0.0799	0.333	0.333	0.333
12	0.2917	0.3162	0.3195	0.3199	1	0	0
20	0.3047	0.3181	0.3198	0.32	0	1	0
21	0.0648	0.0781	0.0798	0.08	0.333	0.333	0.333
22	0.0648	0.0781	0.0798	0.08	0.333	0.333	0.333

Table 14.2: Estimated state value function for paper–rock–scissors problem. Column label represents number of iterations. Final three columns represent the policy π^3 obtained after four iterations.

The final results are shown in Table 14.2. We see that the value-iteration algorithm converged (with tolerance $\epsilon = 0.001$ and discount factor $\delta = 0.5$) after only four rounds. The final policy is also given in the table, which agrees with our conclusion of Example 14.2.   ▷

14.3 From Monte Carlo to Temporal Differences

Previously, we saw how dynamic programming (DP) methods can be used to solve reinforcement-learning problems when the underlying dynamics of the system are known. In the remainder of the chapter, we will discuss *learning* algorithms, that task our intrepid agent with finding an optimal policy by continuously interacting with an a priori unknown environment.

The primary focus of this chapter is on *temporal-difference* (TD) methods, which lie at the heart of much of modern-day reinforcement learning. Before discussing these techniques, however, we will start our discussion by laying out a few definitions regarding the distinction between *on-policy* and *off-policy* learning. We then show how one can solve for the optimal policy in the context of episodic tasks using Monte-Carlo methods. Only then will we get into the heart of the section as we build up the theory of temporal-difference methods.

14.3.1 Learning Through Experience

Before diving into temporal-difference learning, we begin with a few general words to highlight our departure from learning from a completely known model to learning from experience.

Generalized Policy Iteration

During our discussion on dynamic programming, we discussed *policy iteration*, which is an iterative technique that alternates between a policy evaluation (prediction) step and a policy improvement (update) step. We will continue to follow the this process, which is known as *generalized policy iteration*, for the learning problem. Moreover, as was the case in the value iteration algorithm, we will tend to favor algorithms that perform both steps within a given algorithm. To achieve this, we will maintain both an approximate value function as well as an approximate policy. With each step, the value function gets closer to the current policy, and the policy is improved towards becoming the optimal policy.

The crucial difference between general reinforcement learning methods and dynamic programming is, of course, the lack of knowledge of the underlying model. In the process of dynamic programming, we never had to learn the state-action value function directly, as we could simply compute it given our estimate of the state-value function V and the underlying transition dynamics φ, as in Equation (14.21). As a result, general reinforcement learning algorithms focus on learning the state-action value function Q, as opposed to the state value function V. We can then use Equation (14.18) to find a more optimal policy, as opposed to the simplified form Equation (14.20) that can be used when the transition dynamics are known.

On-Policy and Off-Policy Exploration

Exploration–Exploitation Tradeoff

Learning state-action value functions presents an additional challenge compared with learning state value functions, since a value for each state *and* each possible action must be learned through experience. If we simply follow the locally greedy policy, it is likely that we each state will harbor a set of actions that were never selected. This is counterproductive to the task of learning, since the act of only trying what you already think to be best is antithetical to the necessary step of trying new things. This is known as the *exploration–exploitation tradeoff*, which we first introduced during our discussion of multi-armed bandits in Section 4.5. Successful reinforcement learning algorithms must balance the need to find an optimal solution with the need to explore.

On-Policy and Off-Policy Learning

Exploration and learning can be achieved using either *on-policy* or *off-policy* methods.

Definition 14.13. *An* on-policy *method is a learning algorithm that learns the state-action values of the policy that is used to generate actions.*

An off-policy *method is a learning algorithm that learns the state-action values of one policy, often referred to as the* target *policy, from an agent who acted according to a different policy, often referred to as the* behavioral *policy.*

On-policy learning does not mean the policy is fixed. We can still follow the generalized policy iteration approach, so that the policy slowly evolves over time. However, when we update our estimate of the state-action values, we are using information from the policy that we are following.

Off-policy learning, on the other hand, seeks to learn about one policy while following another. This has two important applications. The first is that we can learn about the optimal policy while following a behavioral policy that maintains exploration. In this approach, the behavioral policy is never an optimal policy relative to our current understanding of the state-action values, as it maintains a certain degree of exploration, which, by definition, involves making non-optimal choices to see if they are actually better than our current optimal estimates. The second application is in learning from others. For example, a reinforcement learning algorithm that is trying to learn how to play a game might learn from many examples of the game behind played by humans. The humans are, of course, doing their own thing, and following their instinct and expertise.

On-Policy Exploration: ϵ-Greedy Policy

A common approach to maintain exploration for on-policy learning is to use an ϵ-greedy strategy (recall Definition 4.20). For our present purposes, we define the ϵ-greedy policy as follows.

Definition 14.14. *Given a Markov decision process $(\mathcal{S}, \mathcal{A}, \varphi, \Upsilon)$ and an arbitrary state-action value function $q(s, a)$, the associated ϵ-greedy policy $\Pi(q; \epsilon)$ relative to q is the policy defined by*

$$\Pi(q; \epsilon)_s = (1 - \epsilon)\text{Unif}(\mathcal{A}_s^q) + \epsilon\text{Unif}(\mathcal{A}_s), \qquad (14.24)$$

where $\epsilon \in [0, 1]$ represents the degree of exploration and $\mathcal{A}_s^q$ is the optimal-action bundle of q given by Equation (14.18).

A typical value of exploration is $\epsilon = 0.1$, though this may be regarded as a tunable hyperparameter.

Essentially, the ϵ-greedy policy chooses a greedy option $(1 - \epsilon)$ of the time, and a random action ϵ of the time. The special case $\epsilon = 0$ corresponds to the greedy policy, whereas $\epsilon = 1$ corresponds to a policy that can never learn, because it is always exploring. For on-policy learning algorithms with nonzero $\epsilon > 0$, we are therefore not learning the optimal policy, but rather learning the the state-action values for near-optimal policies that still maintain a fixed degree of exploration.

Off-Policy Learning and Importance Sampling

Many, though not all, off-policy learning techniques leverage the idea of importance sampling, which is a general technique that can be used to estimate statistics of one distribution using samples from another. (Incidentally, our first off-policy algorithm, Q-learning, will circumvent necessity of this technique.)

Definition 14.15. *Given a Markov decision process $(\mathcal{S}, \mathcal{A}, \varphi, \Upsilon)$, two policies π and b, where b is a soft (i.e., nonvanishing) policy, and a trajectory $\tau = \{(s_{t+k}, a_{t+k})\}_{k=0}^h$ of length h, the importance-sampling ratio ρ of policy π relative to policy b and trajectory τ is given by*

$$\rho = \prod_{k=0}^{h-1} \frac{\pi_{s_{t+k}}(a_{t+k})}{b_{s_{t+k}}(a_{t+k})}. \qquad (14.25)$$

Typically, we interpret the policy b as the behavioral policy, which was used to generate the trajectory τ, and the policy π as the target policy, about which we are seeking to learn. We require the policy b to be soft, meaning nonvanishing, to prevent division by zero in the ratios of Equation (14.25). The importance-sampling ratio is of particular use due to the following result.

Proposition 14.4. *Given a Markov decision process $(\mathcal{S}, \mathcal{A}, \varphi, \Upsilon)$, two policies π and b, where b is a soft policy, and a trajectory $\tau = \{(s_{t+k}, a_{t+k})\}_{k=0}^h$ of length h, the importance sampling ratio ρ represents the relative probability of the trajectory occurring under policy π relative to policy b; i.e.,*

$$\rho = \frac{\mathbb{P}_\pi(\tau | S_t = s_t)}{\mathbb{P}_b(\tau | S_t = s_t)}. \qquad (14.26)$$

Note that we implicitly disregard the final action $A_{t+h} = a_{t+h}$, as we are concerned with the probability of reaching the final state $S_{t+h} = s_{t+h}$ under the set of intermediate actions.

Proof. The probability of the trajectory τ relative to policy π is given by

$$\mathbb{P}_\pi(\tau|S_t = s_t) = \pi_{s_t}(a_t)\varphi_{s_t}^{a_t}(s_{t+1})\cdots\pi_{s_{t+h-1}}(a_{t+h-1})\varphi_{s_{t+h-1}}^{a_{t+h-1}}(s_{t+h})$$

$$= \prod_{k=0}^{h-1}\pi_{s_{t+k}}(a_{t+k})\varphi_{s_{t+k}}^{a_{t+k}}(s_{t+k+1}).$$

Therefore, the relative probability of the trajectory τ occurring under policy π relative to policy b is given by

$$\frac{\mathbb{P}_\pi(\tau|S_t = s_t)}{\mathbb{P}_b(\tau|S_t = s_t)} = \frac{\prod_{k=0}^{h-1}\pi_{s_{t+k}}(a_{t+k})\varphi_{s_{t+k}}^{a_{t+k}}(s_{t+k+1})}{\prod_{k=0}^{h-1}b_{s_{t+k}}(a_{t+k})\varphi_{s_{t+k}}^{a_{t+k}}(s_{t+k+1})}.$$

However, the state-transition probabilities in the numerator and denominator cancel, so that the preceding equation reduces to the importance-sampling ratio given by Equation (14.25). This proves the result.   □

14.3.2 Monte-Carlo Control

A simple method for solving reinforcement-learning problems is by using Monte-Carlo methods, which essentially attempt to memorize the expected returns across the entire state-action space. In the context of reinforcement learning, Monte-Carlo methods are only suitable for episodic tasks. For Monte-Carlo control, which seeks to learn an unknown optimal policy, the idea is that one waits for the end of each episode and then updates the state-action value function for each state-action pair encountered along the path through the episode. A similar technique can also be used for the policy-evaluation problem; we leave the details to Sutton and Barto [2018]. The Monte-Carlo solution to the control problem is given in Algorithm 14.5.

Monte-Carlo methods are often used with a concept known as *exploring starts*, as defined below.

Definition 14.16 (Exploring Starts). *Given an episodic MDP, exploring starts is any exploration strategy that satisfies*

1. *the first state-action pair of each episode is selected at random, such that each state-action pair has a nonzero probability of being selected;*
2. *subsequent actions are greedy relative to the current estimated state-action value function Q.*

The initial action for each episode, shown on line 6, is exploratory, whereas subsequent actions, shown on line 14, are greedy. In addition to Q, the Monte-Carlo control problem tracks an array N that represents the number of visits to each state-action pair. The state-action value function is then the average return observed for each state-action pair over many example episodes. Note that line 21 adds the current reward to the cumulative average, for a given state-action pair, and line 23 then averages by dividing the cumulative total reward by the new count.

Unlike dynamic programming, Monte Carlo methods do not require a model, as they operate by aggregating sample experience. They also do not bootstrap; i.e., value estimates do not use other value estimates during the update protocol. They suffer from the drawback that one requires a sufficient number of visits to each state-action pair in order to estimate the value. Such an approach becomes easily untenable for large problems.

We next look at a technique—temporal differencing—that combines the bootstrapping aspects of dynamic programming with the sampling aspects of Monte-Carlo methods. The result is a class of techniques that can not only learn without a model, but also by bootstrapping existing estimates into better ones.

Algorithm 14.5: Monte Carlo Control with Exploring Starts

Input: An MDP $(\mathcal{S}, \mathcal{A}, \varphi, \varrho)$;
      A discount factor $\gamma \in [0,1]$
Output: An optimal state-action value function

1 Initialize an arbitrary function $Q : \mathcal{S} \times \mathcal{A} \to \mathbb{R}$
2 Set $Q(s_0, a) = 0$ for any absorbing state s_0
3 Set $N = \texttt{Zeros}(\mathcal{S} \times \mathcal{A})$
4 **for Each** *episode* **do**
5   Initialize $S \sim \text{Unif}(\mathcal{S})$
6   Select exploratory action $A \sim \Pi(Q;1)_S$
7   Set $path, rewards = \texttt{List}(), \texttt{List}()$
8   **while** S *is not an absorbing state* **do**
9   Observe $S' \sim \varphi_S^A$
10   Observe $R \sim \varrho_{SS'}^A$
11       $\texttt{append}(path, (S, A))$
12       $\texttt{append}(rewards, R)$
13   Update $S = S'$
14   Update by selecting a greedy action $A \sim \Pi(Q;0)_{S'}$
15   **end**
16   Set $G = 0$
17   **while** *path* **is not Empty do**
18   Set $(S, A) = \texttt{pop}(path)$
19   Update $G = \gamma G + \texttt{pop}(rewards)$
20   **if** (S, A) **not in** *path* **then**
21   Update $Q[S, A] = N[S, A] * Q[S, A] + G$
22   Update $N[S, A] = N[S, A] + 1$
23   Update $Q[S, A] = Q[S, A]/N[S, A]$
24   **end**
25   **end**
26 **end**
27 **return** Q

14.3.3 TD Policy Evaluation

Temporal-difference learning plays a central role in modern reinforcement learning. It is not dissimilar to gradient descent, in the sense that it makes small correctional steps for each iteration of the algorithm, with the step size being controlled by a parameter η corresponding to the learning rate. Though not used in the control problem, we begin with an illustration of temporal-difference learning in the context of policy evaluation. (For the control problem, we require estimation of the state-action value function.)

Essentially, temporal-difference methods combine sampling and bootstrapping to form updates of the value function. In this section, we will focus on *one-step* temporal-difference methods, but will generalize to *multi-step* methods thereafter. At heart of the policy update is the *TD error*, defined by

$$\delta V_t = R_{t+1} + \gamma V(S_{t+1}) - V(S_t). \tag{14.27}$$

This represents the difference between the value estimate at step t, $V(S_t)$, and the (presumably better) value estimate that combines the subsequent reward with the value of the transition state; i.e., $R_{t+1} + \gamma V(S_{t+1})$. Note that the temporal-difference error is not immediately available, but rather it is available only after we have observed the imminent reward and transition state.

Algorithm 14.6: One-step TD policy evaluation.

Input: An MDP $(\mathcal{S}, \mathcal{A}, \varphi, \varrho)$;
    A given policy π;
    An initial guess (optional) $V : \mathcal{S} \to \mathbb{R}$;
    A discount factor $\gamma \in [0, 1]$;
    A learning rate $\eta \in (0, 1]$
Output: An estimate of the value function v_π of the given policy

1   **if** V **is Null then**
2   Initialize an arbitrary function $V : \mathcal{S} \to \mathbb{R}$
3   Set $V(s_0) = 0$ for any absorbing state s_0
4   **end**
5   **for Each** *episode* **do**
6   Initialize $S \sim \mathrm{Unif}(\mathcal{S})$
7   **while** S *is not an absorbing state* **do**
8   Select action $A \sim \pi_S$
9   Observe $S' \sim \varphi_S^A$
10   Observe $R \sim \varrho_{SS'}^A$
11   Update $V(S) = V(S) + \eta\,[R + \gamma V(S') - V(S)]$
12   Update $S = S'$
13   **end**
14   **end**
15   **return** V

The temporal-difference method for policy evaluation is shown in Algorithm 14.6. Note that unlike the dynamic-programming policy evaluation algorithm (Algorithm 14.2), the TD method learns with experience in real time. As such, we explicitly reference the various episodes for the case of episodic tasks. For continuing tasks, we may disregard this outer for loop.

The TD policy evaluation method is not a control algorithm, as the policy π remains fixed. It is simply a method for updating state values of the policy as the consequences of the policy unfold. Further, we need not know the true state and reward dynamics φ and ϱ. We only require the *outcome* S' and R of those dynamics to proceed. Thus, the algorithm can learn from *actual* experience, without regard to an a priori understanding of the system dynamics. Moreover, the algorithm can further learn from *simulated* experience, though we must prescribe a (often simple) model of those dynamics to run simulations. For the purpose of simulation, the model only need generate sample transitions; a full specification of the underlying probability density, as would be required for dynamic programming, is unnecessary.

14.3.4 On-Policy TD Control: SARSA

As we stated at the top of section, learning the state value function for experiential learning algorithms is totally insufficient, as we have no knowledge of the underlying system dynamics to determine a greedy policy from our state-value estimates. The temporal-difference control algorithm therefore necessarily involves estimating the state-action values $Q(S, A)$. In fact, there is a natural generalization of the temporal-difference error of Equation (14.27) to the state-action value estimates. The resulting algorithm, known as *SARSA*, is provided in Algorithm 14.7.

Note that SARSA replaces the state-based TD error of Equation (14.27) with the state-action alternative

$$\delta Q_t = R_{t+1} + \gamma Q(S_{t+1}, A_{t+1}) - Q(S_t, A_t). \qquad (14.28)$$

The quantity $Q(S_t, A_t)$ represents the *current* estimate for the value of the current state-action pair, whereas the quantity $R_{t+1} + \gamma Q(S_{t+1}, A_{t+1})$ represents a better value as based on the immediately observed return and the discounted estimate of the subsequent state-action pair. Each update therefore requires the quintuple $(S_t, A_t, R_{t+1}, S_{t+1}, A_{t+1})$, from which the SARSA algorithm derives its name.

The SARSA algorithm further selects actions relative to the corresponding ϵ-greedy policy $\Pi(Q; \epsilon)$, which we previously defined in Equation (14.24). SARSA therefore is learning a suboptimal policy that maintains a certain degree of exploration. Naturally, as $\epsilon \to 0$, the algorithm converges on the optimal policy, assuming that it has sufficiently sampled the state-action bundle $\mathcal{A}_s$ consisting of all admissible state-action pairs.

14.3.5 Off-Policy TD Control: Q Learning

Next, we consider a simple yet powerful modification to SARSA that yields an elegant off-policy one-step temporal-difference control algorithm called *Q learning* (Watkins [1989] and Watkins and Dayan [1992]), as shown in Algorithm 14.8.

The key difference between SARSA and Q learning is that the latter uses the difference

$$\delta Q_t^* = R_{t+1} + \gamma \max_{a \in \mathcal{A}_{S_{t+1}}} Q(S_{t+1}, a) - Q(S_t, A_t) \tag{14.29}$$

in place of Equation (14.28). This seemingly innocuous change has a profound result: the method now learns the *optimal policy*, regardless of the actual policy being followed. This is due to the fact that the update rule compares the current state-action value with the immediate return and the discounted value of the optimal value at the subsequent state.

The backup diagrams for both SARSA and Q learning are shown in Figure 14.5. SARSA uses the actual values of the subsequent state and action to update its current state-action value estimates. Q learning, on

---

Algorithm 14.7: SARSA: One-step On-Policy TD control.

Input: An MDP $(\mathcal{S}, \mathcal{A}, \varphi, \varrho)$;
  An initial state-action value function (optional) $Q : \mathcal{S} \times \mathcal{A} \to \mathbb{R}$;
  A discount factor $\gamma \in [0, 1]$;
  A learning rate $\eta \in (0, 1]$;
  An exploration parameter $\epsilon \in (0, 1)$
Output: An optimal state-action value function, given the exploration
  constraint

1 **if** Q **is Null then**
2 | Initialize an arbitrary function $Q : \mathcal{S} \times \mathcal{A} \to \mathbb{R}$
3 | Set $Q(s_0, a) = 0$ for any absorbing state s_0
4 **end**
5 **for Each** *episode* **do**
6 | Initialize $S \sim \text{Unif}(\mathcal{S})$
7 | Select action $A \sim \Pi(Q; \epsilon)_S$
8 | **while** S *is not an absorbing state* **do**
9 | | Observe $S' \sim \varphi_S^A$
10 | | Observe $R \sim \varrho_{SS'}^A$
11 | | Select action $A' \sim \Pi(Q; \epsilon)_{S'}$
12 | | Update $Q(S, A) = Q(S, A) + \eta [R + \gamma Q(S', A') - Q(S, A)]$
13 | | Update $S = S'$
14 | | Update $A = A'$
15 | **end**
16 **end**
17 **return** Q

Algorithm 14.8: Q Learning: One-step Off-Policy TD control.

Input: An MDP $(\mathcal{S}, \mathcal{A}, \varphi, \varrho)$;
An initial state-action value function (optional) $Q : \mathcal{S} \times \mathcal{A} \to \mathbb{R}$;
A discount factor $\gamma \in [0, 1]$;
A learning rate $\eta \in (0, 1]$;
An exploration parameter $\epsilon \in (0, 1)$

Output: An estimate of the optimal state-action value function q_*

1 **if** Q **is Null then**
2   Initialize an arbitrary function $Q : \mathcal{S} \times \mathcal{A} \to \mathbb{R}$
3   Set $Q(s_0, a) = 0$ for any absorbing state s_0
4 **end**
5 **for Each** *episode* **do**
6   Initialize $S \sim \mathrm{Unif}(\mathcal{S})$
7   **while** S *is not an absorbing state* **do**
8   Select action $A \sim \Pi(Q; \epsilon)_S$
9   Observe $S' \sim \varphi_S^A$
10   Observe $R \sim \varrho_{SS'}^A$
11   Set $Q(S, A) = Q(S, A) + \eta \left[R + \gamma \max_{a \in \mathcal{A}_{S'}} Q(S', a) - Q(S, A) \right]$
12   Set $S = S'$
13   **end**
14 **end**
15 **return** Q

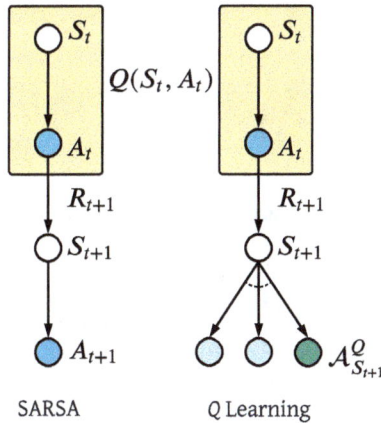

Fig. 14.5: Backup diagram for SARSA (left) and Q learning (right). For Q learning, the greedy action at time $t + 1$ relative to Q is shaded in green; suboptimal (unused) values are shaded a lighter blue.

the other hand, uses the subsequent state, but then looks at all possible actions from the transition state and uses the maximum value. This is why Q learning is off-policy: it uses the best possible action from the transition state to feed back into the state-action value update, regardless of the choice of the actual action that is taken.

14.3.6 Examples

Gridworld Cliffs

Example 14.4 (Gridwolrd). We first introduced Gridworld problems in Example 14.1. In this example, we will solve the gridworld problem in Python using both SARSA and Q learning. This is accomplished with our `Gridworld` class, as shown in Code Block 14.2. The learning algorithm defaults to Q learning, but one may alternatively select SARSA by passing `qlearn=False` into the constructor.

To describe our current state, we will use a coordinate system congruent with `numpy` arrays, so that the first dimension is vertical and oriented downward, and the second dimension is horizontal and oriented to the right. A 5×5 and 4×12 gridworld is shown in Figures 14.4 and 14.6, respectively.

Gridworld problems are simple as they are both deterministic in the reward payout (which only depends on the transition state) and deterministic in the transitions. We will encode our action space as integers using the encoding

$$\text{"down"} \Rightarrow 0 \qquad \text{"right"} \Rightarrow 1 \qquad \text{"up"} \Rightarrow 2 \qquad \text{"left"} \Rightarrow 3.$$

State transitions are encoded in the `getNextState` method on line 11. To handle the boundary, we have tabulated a list of actions, their associated boundary conditions, and the transition rule to generate the subsequent state, if the current state is not along a boundary, in Table 14.3. We make the clever definition of dimension as $a \mod 2$ and sign as $+1$ for $a = 0, 1$ and -1 otherwise.

action	direction	boundary	transition	dimension	sign
$a = 0$	↓	$s(0) = shape(0) - 1$	$i + 1$	0	$+1$
$a = 1$	→	$s(1) = shape(1) - 1$	$j + 1$	1	$+1$
$a = 2$	↑	$s(0) = 0$	$i - 1$	0	-1
$a = 3$	←	$s(1) = 0$	$j - 1$	1	-1

Table 14.3: Gridworld actions, with state $s = (i, j)$.

In order to determine the action, we select the optimal action, relative to our Q matrix, with probability $1 - \epsilon$; otherwise we select an action at random. Finally, our **run** method simulates one entire episode: that state

```
 1  class Gridworld:
 2
 3      def __init__(self, shape=(4,7), rewards=None, start=None,
            absorbers=[], epsilon=0.1, eta=0.1, qlearn=True):
 4          self.shape, self.rewards = shape, rewards
 5          self.start = start if start is not None else (0, 0)
 6          self.absorbers, self.qlearn = absorbers, qlearn
 7          self.epsilon, self.eta = epsilon, eta
 8          self.Q = np.zeros(shape+(4,))
 9          self.returns = []
10
11      def getNextState(self, state, action):
12          dim = action % 2
13          sign = 1 if action < 2 else -1
14          if sign == +1 and state[dim] == self.shape[dim] - 1:
15              return state
16          if sign == -1 and state[dim] == 0:
17              return state
18          next_state = list(state)
19          next_state[dim] += sign
20          return tuple(next_state)
21
22      def getAction(self, state, epsilon=0):
23          Qs = self.Q[state]
24          if np.random.random() < epsilon:
25              actions = np.arange(len(Qs))
26          else:
27              actions = np.argwhere(Qs == max(Qs)).flatten()
28          return np.random.choice(actions)
29
30      def run(self, epsilon=-1):
31          epsilon = self.epsilon if epsilon < 0 else epsilon
32          state = self.start
33          action = self.getAction(state, epsilon=epsilon)
34          self.path = [state]
35          self.returns.append(0)
36          while state not in self.absorbers:
37              next_state = self.getNextState(state, action)
38              reward = self.rewards[next_state]
39              # next_state = self.start if reward < -10 else next_state
40              next_action = self.getAction(next_state, epsilon=epsilon)
41              next_q = max(self.Q[next_state]) if self.qlearn else \
                    self.Q[next_state][next_action]
42              delta = reward + next_q - self.Q[state][action]
43              self.Q[state][action] += self.eta * delta
44              state, action = next_state, next_action
45              self.path.append(state)
46              self.returns[-1] += reward
```

Code Block 14.2: Gridworld class

is initialized to the starting state, and then updated until the state lands on an absorber. δQ_t is determined on line 42, where `next_q` is determined on the preceding line, based on whether the desired algorithm is SARSA or Q learning.

▷

Example 14.5 (Gridworld Cliffs). Our next example, due to Sutton and Barto [2018], nicely illustrates the difference between the on-policy SARSA and off-policy Q learning algorithms. This is similar to the gridworld problem of Example 14.1, since each transition carries a penalty of $R = -1$. The difference, however, is there is a segment, shaded in red, that carries a high penalty of $R = -100^2$.

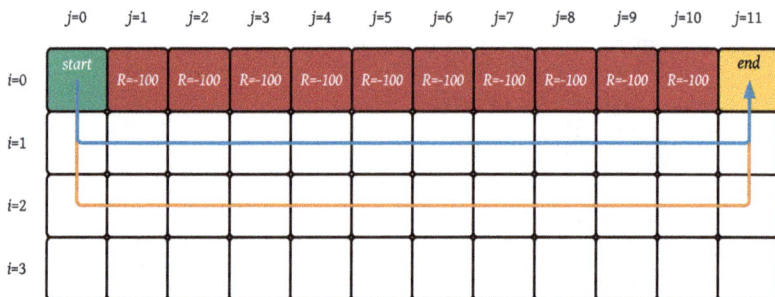

Fig. 14.6: Gridworld Cliffs example. Red region represents cliff, with penalty $R = -100$. Blue line: Q learning. Orange line: SARSA.

The optimal path is obviously the blue path that proceeds down row $i = 1$. This is the policy learned by Q learning. SARSA, on the other hand, learns the orange path, which proceeds down row $j = 2$. By increasing the penalty of the cliff, one can even push the SARSA path further out, into row $j = 3^3$. We can simulate the gridworld cliffs example using our `Gridworld` class, as shown in Code Block 14.3.

Note that the total return is calculated for each episode. We can further use a moving-average to smooth out some of the noise, as shown on line 10. The MA 10 and MA 100 smoothed returns for Q learning and SARSA are shown in Figure 14.7.

When following the greedy path, Q learning will typically beat SARSA. After all, it is learning the optimal path, while following a path that main-

---

[2] In Sutton and Barto [2018], landing on the red squares also reset the state back to the start. This can be achieved by uncommenting line 39 in Code Block 14.2.

[3] The far path, along row $j = 3$, is actually the SARSA solution illustrated in Sutton and Barto [2018].

```
1  depth, width = 4, 12
2  rewards = -np.ones((depth, width))
3  rewards[0, 1:-1] = -100
4  GS = Gridworld(shape=(depth, width), rewards=rewards,
       absorbers=[(0,width-1)], qlearn=False)
5  G = Gridworld(shape=(depth, width), rewards=rewards,
       absorbers=[(0,width-1)])
6  for i in range(1000):
7      G.run()
8      GS.run()
9  # example smoothing
10 np.convolve(G.returns, np.ones(10)/10, mode='valid')
```

Code Block 14.3: Gridworld-cliffs example.

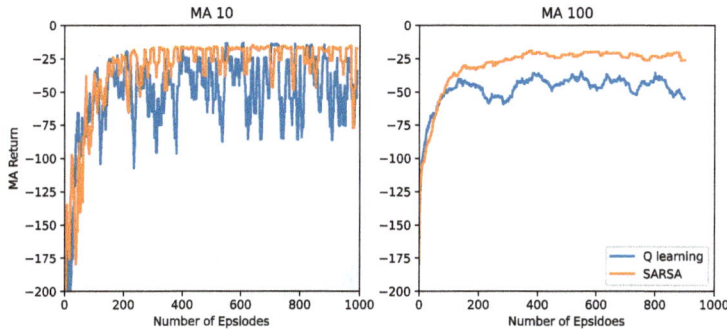

Fig. 14.7: Smoothed moving average total return for Q learning (blue) and SARSA (orange).

tains exploration. Exploratory steps, however, can easily result in a stiff penalty, due to the fact that Q learning's greedy path traverses the edge of the cliff. SARSA, on the other hand, learns based on its actual experience; i.e., it is online. Due to the exploratory steps, it actually learns to *stay away* from the cliff, as it learns that walking along the cliff can lead to a costly mistake.

Finally, we can actually plot the state-action value function, as shown in Code Block 14.4[4]. The resulting plots are shown in Figure 14.8. Note due to the inversion of the y-axis, the actions 0 and 2 now represent up and down, respectively.

Figure 14.8 makes a good deal of sense. Starting from state $(0,0)$, moving up $(a = 0)$ is only marginally better than moving down $(a = 2)$ or left $(a = 3)$, as these latter two options incur a modest cost of $R = -1$ and do

---

[4] We ran an additional 10,000 iterations to get convergence.

```
1  ## Plot state-action value function
2  x = np.arange(0, 13)
3  y = np.arange(0,5)
4  X_grid, Y_grid = np.meshgrid(x, y)
5  cm = colors.LinearSegmentedColormap.from_list('mylist',
       ['#d62728','#1f77b4'], N=2)
6  fig, ax = plt.subplots(2, 2, figsize=(10, 7))
7  for i in range(2):
8      for j in range(2):
9          im = ax[i, j].pcolormesh(X_grid, Y_grid, G.Q[:, :, 2*i+j],
               edgecolors='white', vmin=-25, vmax=0)
10         ax[i,j].set_xticks(np.arange(12)+0.5)
11         ax[i,j].set_xticklabels(np.arange(12))
12         ax[i,j].set_yticks(np.arange(4)+0.5)
13         ax[i,j].set_yticklabels(np.arange(4))
14         ax[i,j].set_title(f"Action = {2*i+j}")
15  fig.colorbar(im, ax=ax.ravel().tolist())
```

Code Block 14.4: Gridworld-cliffs example: plotting the state-action value function.

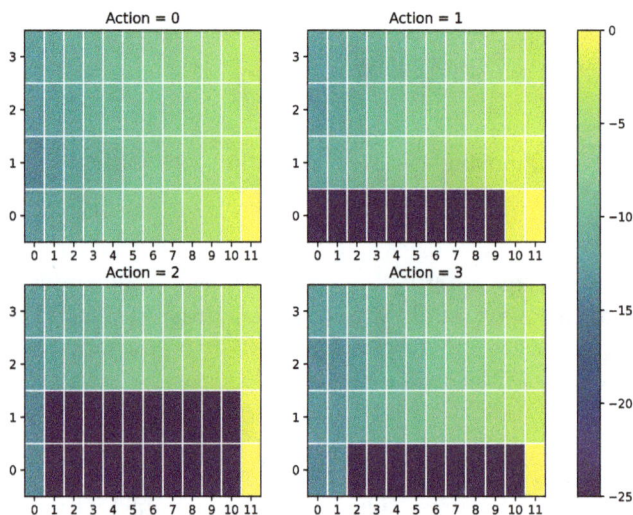

Fig. 14.8: State-value action function for gridworld-cliffs example: 0 represents up, 1 represents right, 2 represents down, and 3 represents left. *Note*: the vertical axis is inverted relative to our earlier diagrams.

not change your state. Moving right ($a = 1$), on the other hand, incurs a penalty of $R = -100$. In general, there is a heavy penalty for any state-action that would result in a transition to the cliff ($i = 0$, $j = 1, \ldots, 11$).

$\triangleright$

Learning *Blackjack*

Example 14.6. Blackjack is a card game between a player (our agent) and a dealer. To begin, two cards are dealt to both the player and the dealer, with one of the dealer's cards being dealt face down and the other face up, as shown in Figure 14.9.

Fig. 14.9: A game of blackjack with state: $(7, 19, 1)$.

The player then either *hits*, thereby receiving an additional card, or *sticks*, thereby ending his turn. Face cards have a value of 10, and the ace may count as either a 1 or 11. The goal is to get as close as possible to 21 without going over (i.e., *going bust*). If the player hits and his value exceeds 21, he is said to *go bust*, and the game ends as a loss (with a reward of -1).

Once the player sticks, assuming he hasn't gone bust, it becomes the dealer's turn. The dealer plays according to the following fixed strategy: the dealer sticks whenever his sum becomes at least 17, otherwise he hits. We will assume an infinite deck, so that cards are dealt with replacement. If the dealer goes bust, the player wins, with a reward of $+1$. Otherwise, once the dealer sticks, the result is determined by who is closest to 21: if the player is closest to 21, the player wins, with a reward of $+1$. If the dealer is closest to 21, the player loses, and receives a reward of -1. Otherwise, if the score of the dealer and player are equal, the game ends as a draw, in which case the reward is $+0$.

Naturally, aces should count as 11 unless it puts the sum over 21, in which case they should be counted as a 1. An ace that is currently counted as an 11 is said to be a *usable ace*, which therefore represents an extra degree of freedom of the game play. Moreover, if the player's sum is less than or equal to 11, the player should always hit, as there is no danger in going bust. The state space for the player (agent) is thereby determined by three numbers: the value of the dealer's face-up card (2–11), the player's current sum (12–21), and whether or not the player has a useable ace (0 or 1). The size of the state space is therefore $|\mathcal{S}| = 200$. The action space is $\mathcal{A} = \{0,1\}$, where 1 is interpreted as a *hit* and 0 is interpreted as a *stick*.

We coded the Q learning algorithm for blackjack in Code Blocks 14.1 and 14.6. The attribute Q represents the learned state-action value function and is constituted as a four-dimensional array with shape $10 \times 10 \times 2 \times 2$. The first three dimensions represent the state (value of dealer's face-up card, value of player's hand, whether player has a useable ace); the fourth dimension represents the two-dimensional action space consisting of 0 for *stick* and 1 for *hit*.

The newGame method initializes an episode. The player and dealer's hands are initialized with two empty lists. Two card are then dealt to both the player and the dealer, using the hit method. If the player's hand has a value less than twelve, the player will always *hit*, as it is impossible to go bust. This is why the slice of state space corresponding to the player's value ranges between 12 and 21. Dealing is accomplished by the hit method, which takes a hand (list object) as an argument. Since the input variable is passed by reference, changes to it affect the upstream variable; this is why there is no *return*. On line 20, the extra [10,10] is appended to the list range(2,12) as a ten-valued card is three times as likely to appear as a non-ten card; i.e., *jack, queen, king*. The *ace* is dealt as an eleven. Next, we have logic that *uses* a *usable ace* if the new total exceeds 21, reducing the current value (11) of the ace to a value of 1.

Next we have the getState method, which encodes the state vector based on the dealer's face up card (first position in the dealer's hand), the total value of the player's hand, and whether or not the player has a usable ace (a card with value 11). The -2 and -12 shifts to the first two dimensions ensure that the state is encoded in the range $\{0,\ldots,9\} \times \{0,\ldots,9\} \times \{0,1\}$, so that it is compatible with the indices of Q, which is an np.array object.

The getAction method returns the locally greedy strategy $1 - \epsilon$ of the time and a random strategy ϵ of the time, virtually identical to its functioning in the Gridworld example of Code Block 14.2.

The playGame method simulates a single episode. First, it initializes a new game. Then it loops over the player's decisions until the player either decides to stick or goes bust. Since the reward is only paid out at the end, there is no immediate reward during the *while loop*; rather the update relies on propagating the optimal state-action value estimate at the transition state back to the current state. Once the *while loop* is complete, the state-action value function is updated with the final reward (win, lose, or draw) at the penultimate state. Since the final state is an implicit absorbing state,

```python
class Blackjack:

    def __init__(self, eta=0.1, epsilon=0.1, gamma=1, lambda_=0):
        self.eta = eta # learning rate
        self.epsilon = epsilon # exploration rate
        self.gamma = gamma # discount factor
        self.lambda_ = lambda_ # trace decy / used later
        self.Q = np.zeros((10,10,2,2)) # 3d state, 1d action
        self.n_games = 0 # counter

    def newGame(self):
        self.dealer = []
        self.player = []
        for i in range(2):
            self.hit(self.dealer)
            self.hit(self.player)
        while sum(self.player) < 12:
            self.hit(self.player)

    def hit(self, hand):
        new_card = np.random.choice([i for i in range(2, 12)] +
            [10,10,10])
        hand.append(new_card)
        if sum(hand) > 21 and 11 in hand:
            ace = np.argmax(hand)
            hand[ace] = 1

    def getState(self):
        return (self.dealer[1]-2, sum(self.player)-12, int(11 in
            self.player))

    def getAction(self, state, epsilon=0):
        Qs = self.Q[state]
        if np.random.random() < epsilon:
            actions = np.arange(len(Qs))
        else:
            actions = np.argwhere(Qs == max(Qs)).flatten()
        return np.random.choice(actions)
    # continued...
```

Code Block 14.5: Blackjack Q learning.

```
1    # continued...
2    def playGame(self):
3        self.newGame()
4        go = True
5        self.n_games += 1
6        state = self.getState()
7        action = self.getAction(state, epsilon=self.epsilon)
8        while go:
9            if action == 1:
10               self.hit(self.player)
11           if (action == 0) or (sum(self.player) > 21):
12               go = False
13           if go:
14               next_state = self.getState()
15               delta = self.gamma * max(self.Q[next_state]) -
                   self.Q[state][action]
16               self.Q[state][action] += self.eta * delta
17               state = next_state
18               action = self.getAction(state, epsilon=self.epsilon)
19       delta = self.getReward() - self.Q[state][action]
20       self.Q[state][action] += self.eta * delta
21
22   def getReward(self):
23       if sum(self.player) > 21:
24           return -1
25       while sum(self.dealer) < 17:
26           self.hit(self.dealer)
27       return np.sign(sum(self.player)-sum(self.dealer)) if
           sum(self.dealer) <= 21 else 1
```

Code Block 14.6: Blackjack Q learning (continued).

there is no contribution due to any transition state. The getReward method then issues a reward of -1 if the player went bust; otherwise it simulates the draw's from the dealer's mandatory policy and then determines the final result.

Now that we have our Blackjack class encoded, we can simulate a number of games, as shown on the first few lines of Code Block 14.7. Notice our optimal use of the tqdm wrapper from the tqdm module, which prints a status bar for the loop, showing the percentage complete along with an estimated time. The code block further provides all necessary code to plot the current greedy policy along with the state-value action functions, the results of which are shown in Figure 14.10 and Figure 14.11.

The strategy shown in Figure 14.10 is *not* the *Baldwin strategy* (also known as the *basic strategy*), which constitutes the optimal strategy for the given blackjack problem (see Example 10.3, Baldwin, *et al.* [1956], and

```python
# Blackjack Simulation -- Q Learning
B = Blackjack(eta=0.015, epsilon=0.15)
for i in tqdm(range(1000000)):
    B.playGame()

# plot policy based on Q
fig, ax = plt.subplots(1, 2, figsize=(10, 4))
x = np.arange(2, 13)
y = np.arange(12,23)
X_grid, Y_grid = np.meshgrid(x, y)
cm = colors.LinearSegmentedColormap.from_list('mylist',
    ['#d62728','#1f77b4'], N=2)
for j in range(2):
    A = np.argmax(B.Q[:, :, j, :], axis=2)
    im = ax[j].pcolormesh(X_grid, Y_grid, A.T, cmap=cm,
        edgecolors='white', vmin=-0.8, vmax=0.8)
    ax[j].set_title(f"Usable Ace = {bool(j)}")
    ax[j].set_xlabel("Value of Dealer's Face-up Card")
    ax[j].set_xticks(np.arange(2,12)+0.5)
    ax[j].set_xticklabels(np.arange(2,12))
    ax[j].set_yticks(np.arange(12,22)+0.5)
    ax[j].set_yticklabels(np.arange(12,22))
ax[0].set_ylabel("Value of Agent's Hand")

# plot state-action value function Q
fig, ax = plt.subplots(2, 2, figsize=(10, 7))
for i in range(2):
    for j in range(2):
        im = ax[i, j].pcolormesh(X_grid, Y_grid, B.Q[:, :, j, i].T,
            edgecolors='white', vmin=-0.8, vmax=0.8)
        ax[i,j].set_xticks(np.arange(2,12)+0.5)
        ax[i,j].set_xticklabels(np.arange(2,12))
        ax[i,j].set_yticks(np.arange(12,22)+0.5)
        ax[i,j].set_yticklabels(np.arange(12,22))
ax[0,0].set_title('Useable Ace = False')
ax[0,1].set_title('Useable Ace = True')
ax[0,0].set_ylabel('A=Stick')
ax[1,0].set_ylabel('A=Hit')
fig.colorbar(im, ax=ax.ravel().tolist())
```

Code Block 14.7: Blackjack simulation and plots

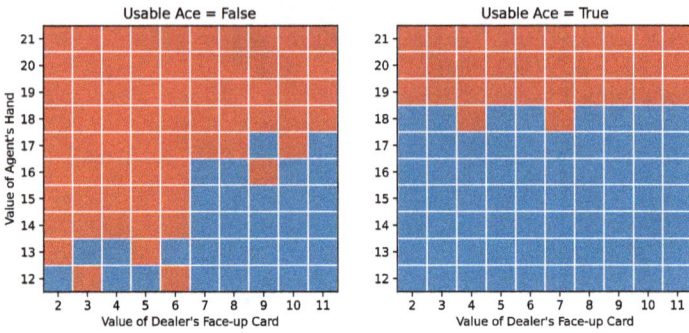

Fig. 14.10: Greedy strategy relative to learned Q for blackjack after 10,000,000 rounds of Q learning: hit (red) or stick (blue).

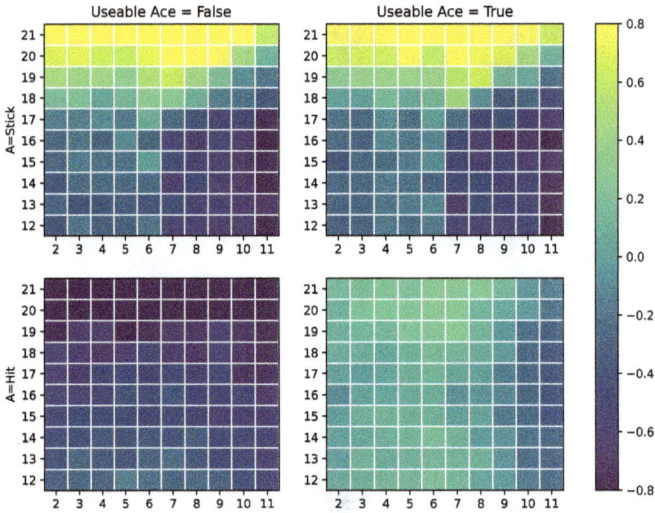

Fig. 14.11: State-action value function Q for for blackjack after 10,000,000 rounds of Q learning.

Thorp [1966]). (The Baldwin strategy further accounts for the options of *splitting* and *doubling down*, which we omit for simplicity.) We further modified our code to replace our ϵ-greedy exploration strategy with exploring starts, only to achieve similar results. It is a bit unsettling that we did not converge on the optimal strategy using Q learning and 10 million simulated games. In our next example, we try our luck using the Monte-Carlo approach, which was used in Sutton and Barto [2018] to determine the optimal strategy.   ▷

Example 14.7. Instead of Q learning, let us use the Monte-Carlo control algorithm to solve the blackjack problem. We can subclass our existing `Blackjack` class, overwriting the learning method within the `playGame` method, as shown in Code Block 14.8. Furthermore, we modify our previous algorithm to use exploring starts.

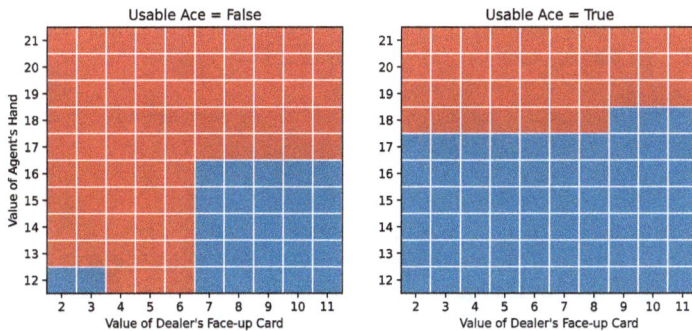

Fig. 14.12: Optimal blackjack policy, obtained using Monte Carlo with exploring starts after 1,000,000 rounds: hit (red) or stick (blue).

Exploring starts is implements with the modification to lines 10 and 21: the first action selection of each episode is exploratory whereas subsequent action selections are greedy. In order to implement the MC method, we track a new attribute `n_visits` that tabulates the number of visits to each state-action pair. Within each episode, then, visits to the state-action pairs are tracked with the `path` variable, which is a list of state-action pair tuples. Once each episode is complete and the reward determined, we then trace backwards through the path, decaying the reward using the discount parameter γ and adding the result to the running average of the state-action value estimate. The backwards loop is achieved by *popping* the final state-action pair of the `path` list.

Further note that a given state-action value is only updated for the *first* occurrence of the pair in each episode's path. This is a standard implementation, though, in blackjack, there are no cycles so the exception is mostly decorative.

```
1   class BlackjackMC(Blackjack):
2
3       def playGame(self):
4           if self.n_games == 0:
5               self.n_visits = np.zeros((10,10,2,2))
6           self.newGame()
7           go = True
8           self.n_games += 1
9           state = self.getState()
10          action = self.getAction(state, epsilon=1) # Exploring Start
11          path =[]
12          while go:
13              path.append(state + (action,))
14              if action == 1:
15                  self.hit(self.player)
16              if (action == 0) or (sum(self.player) > 21):
17                  go = False
18              if go:
19                  next_state = self.getState()
20                  state = next_state
21                  action = self.getAction(state, epsilon=0)
22          reward = self.getReward()
23          while path:
24              state_action = path.pop()
25              if state_action not in path:
26                  self.Q[state_action] = self.n_visits[state_action] *
                        self.Q[state_action] + reward
27                  self.n_visits[state_action] += 1
28                  self.Q[state_action] /= self.n_visits[state_action]
29              reward *= self.gamma
```

Code Block 14.8: `BlackjackMC` Monte-Carlo Blackjack with exploring starts

We ran the MC blackjack method with a discount of $\gamma = 1$ one million times. The result is shown in Figure 14.12. The corresponding estimated state-action value function is shown in Figure 14.13. This represents the true optimal strategy (the Baldwin strategy), in agreement with Sutton and Barto [2018].   ▷

There are two primary differences between the one-step Q-learning and the MC solutions to the blackjack problem. The first is that the one-step method can only propagate the final reward backwards one step at a time, whereas MC was more holistic in that rewards percolate through all past experience. The second is that the TD method uses a constant step size, whereas the MC method uses a step size that decays inversely proportionally to the number of visits to each state-action pair (see Exercise 14.4).

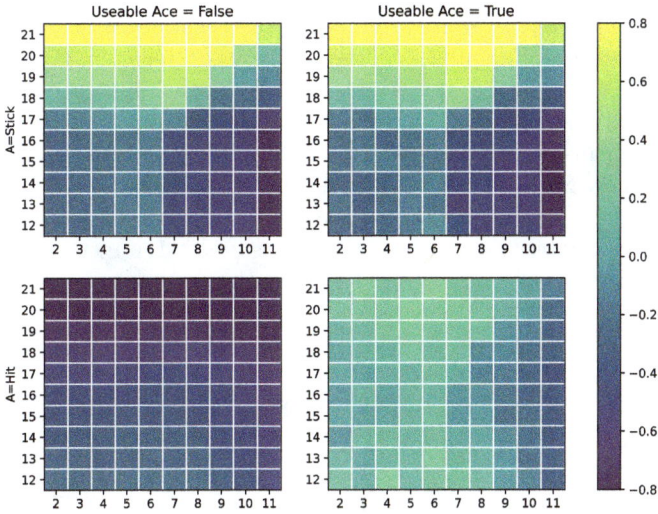

Fig. 14.13: Estimated optimal state-action value function obtained using Monte Carlo with exploring starts after 1,000,000 rounds.

14.3.7 Boosted Starts

Another key to improving the Q-learning algorithm is to recognize that the locally optimal policy differs from the true optimal policy exactly for the states that have the smallest absolute difference $|Q(s,1) - Q(s,0)|$ between their action values (see Exercise 14.6). This motivates the following definition, which we will implement in Code Block 14.9.

Definition 14.17 (Boosted Starts). *Given an episodic MDP,* boosted starts *with* offset parameter ε *is a particular exploring starts strategy that selects an initial state relative to*

$$\mathbb{P}(S_0 = s) \propto \left(\varepsilon + \left|\max_{a \in \mathcal{A}_s} Q(s,a) - \max_{a \neq a^*} Q(s,a)\right|\right)^{-1}, \qquad (14.30)$$

where $a^* = \arg\max_a Q(s,a)$, *and then a corresponding initial action relative to* $A_0 \sim \mathrm{Unif}(\mathcal{A}_{S_0})$.

Recall that $\mathcal{A}_s^Q$ is the arg max of Q over a, so that the absolute difference is between the maximum and penultimate maximum values of Q. A typical value of the offset might be $\varepsilon = 0.01$.

14.4 Eligibility Traces

One obvious shortcoming of one-step temporal-difference methods is that observed rewards only update the immediately preceding step. For episodic tasks in which the reward is not paid out until the end, such as two player games, this can add significant inefficiency to the learning process.

The goal of this section is to generalize the one-step temporal-difference methods to allow for learning across multiple steps. There are two primary methods of achieving this: multi-step temporal-difference methods and eligibility traces. Eligibility traces are themselves multi-step methods, except that they a decay parameter λ, that decays the impact of the reward as the number of intermediary steps increases.

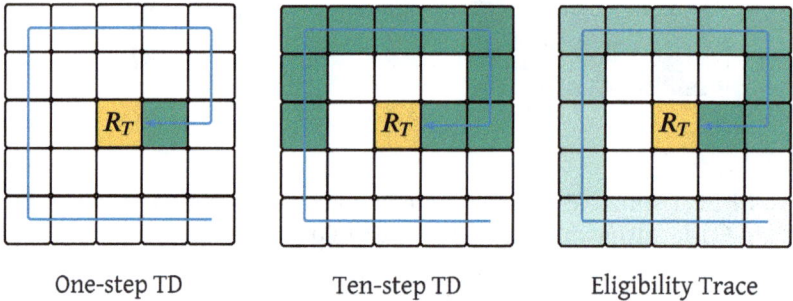

| One-step TD | Ten-step TD | Eligibility Trace |

Fig. 14.14: Learning from the reward R_T: one-step TD, ten-step TD, and eligibility trace methods. Updated states are shaded in green; lighter shades indicate reduced impact.

To illustrate this distinction, consider the gridworld example shown in Figure 14.14. We have a length-17 path that starts in the lower-right corner and terminates in the center (gold) square that yields a reward payout. For a one-step temporal-difference method, only the penultimate state value is updated, as shown in Figure 14.14 (left). For a ten-step method, the ten preceding steps are updated, as shown in Figure 14.14 (center). The reward R_T is decayed according to the discount factor and time, but otherwise no decay is applied. Finally, an eligibility trace is shown in Figure 14.14 (right), in which the reward decays according to a *trace-decay* parameter $\lambda \in [0, 1]$.

Therefore, both multi-step and eligibility-trace methods provide a continuum of techniques that start with one-step TD methods at one end of the spectrum. At the other end of the spectrum, we might consider infinite-step TD methods, or an eligibility trace with $\lambda = 1$ (indicating no decay). Such methods are known as *Monte-Carlo* methods and are treated extensively in Sutton and Barto [2018]. Monte-Carlo methods are only applicable for episodic tasks, and they allow the final return at the *end* of each episode to trickle back over the sequence of states realized for the given episode. They suffer from two downsides: the method cannot learn until each episode is complete and, due to the infinite horizon, they cannot be used for continuing tasks.

14.4.1 n-step Temporal-Difference Methods

To generalize the one-step temporal-difference methods to multi-step methods, we first need to define the n-step return. We will then briefly discuss the n-step TD prediction and control problems to highlight their differences; the full algorithms are given in Sutton and Barto [2018].

The n-step Return

Consider again the TD error defined in Equation (14.27). This error can be represented as the difference between the *one-step return*

$$G_{t,1} = R_{t+1}$$

plus the discounted transition-state value $\gamma V(S_{t+1})$ and the current state value $V(S_t)$; i.e.,

$$\delta V_t = (G_{t,1} + V(S_{t+1})) - V(S_t).$$

We thus see that the TD error is easily generalized to multiple step methods. For instance, to incorporate a second step into our update, we could replace $G_{t,1}$ with a *two-step return*

$$G_{t,2} = R_{t+1} + \gamma R_{t+2},$$

and the discounted transition value $\gamma V(S_{t+1})$ with $\gamma^2 V(S_{t+2})$, thereby obtaining the *two-step TD error*

$$\begin{aligned}
\delta V_{t,2} &= \left[R_{t+1} + \gamma R_{t+2} + \gamma^2 V(S_{t+2}) \right] - V(S_t) \\
&= \left[G_{t,2} + \gamma^2 V(S_{t+2}) \right] - V(S_t).
\end{aligned}$$

Following the same pattern, we could further obtain a *three-step TD error*

$$\begin{aligned}
\delta V_{t,3} &= \left[R_{t+1} + \gamma R_{t+2} + \gamma^2 R_{t+3} + \gamma^3 V(S_{t+3}) \right] - V(S_t) \\
&= \left[G_{t,3} + \gamma^3 V(S_{t+3}) \right] - V(S_t),
\end{aligned}$$

and so forth. Each of these represents the difference between the value estimate of the current state and a better value estimate based on a number of observed subsequent returns and the value estimate of the final transition state.

Note 14.2. Sutton and Barto [2018] defines the *n-step returns* inclusive of the value of their respective transition states, but then redefines the meaning when discussing the *n*-step SARSA algorithm to shift from state values to state-action values. We follow a different definition for two reasons: the phrase *n-step return* implies, to us, the actual (discounted) returns from the first *n*-step and our definition/notation does not change meaning as we move from state value functions to state-action value functions.   ▷

Definition 14.18. *The n-step return of a Markov decision process at time t is nth partial sum of the full return at time t (Definition 14.8), as given by*

$$G_{t,n} = R_{t+1} + \gamma R_{t+2} + \cdots + \gamma^{n-1} R_{t+n} = \sum_{k=0}^{n-1} \gamma^k R_{t+k+1}; \qquad (14.31)$$

i.e., the n-step return is the discounted sum of the n subsequent rewards.

Given this definition, we next consider how to generalize our one-step value estimates.

Definition 14.19. *The n-step state and state-action value estimates of a Markov decision process at time t are given by the equations*

$$\hat{V}_{t,n} = G_{t,n} + \gamma^n V(S_{t+n}) \qquad (14.32)$$
$$\hat{Q}_{t,n} = G_{t,n} + \gamma^n Q(S_{t+n}, A_{t+n}), \qquad (14.33)$$

respectively. Similarly, the n-step state and state-action TD errors are defined by the relations

$$\delta V_{t,n} = \hat{V}_{t,n} - V(S_t) \qquad (14.34)$$
$$\delta Q_{t,n} = \hat{Q}_{t,n} - Q(S_t, A_t), \qquad (14.35)$$

respectively.

The *n*-step value estimates represent improvements on the estimates $V(S_t)$ and $Q(S_t, A_t)$ that incorporate information obtained from the subsequent *n* steps.

n-step TD Methods

n-step temporal-difference learning is a natural extension of its single-step counterpart and is designed to leverage multiple returns for each update. They are based on the difference between the *n*-step state and state-action value estimates (Equations (14.32) and (14.33)) and the current value estimates. One drawback is that we must wait *n* steps before we can apply each update, creating a lag in the learning that must "catch up" at the end of each episode. Since our primary focus of this section is on eligibility traces, which do not suffer from this drawback and otherwise offer certain computational advantages, we will only provide a cursory introduction to multi-step methods, referring the reader to Sutton and Barto [2018] for additional details.

n-step Policy Evaluation

The n-step policy evaluation method generalizes Algorithm 14.6 to include the n-step return. Fundamentally, this is achieved by replacing the one-step TD error with its n-step equivalent, given by Equation (14.34), which may be expressed in terms of the n-step return as

$$\delta V_{t,n} = [G_{t,n} + \gamma^n V(S_{t+n})] - V(S_t), \tag{14.36}$$

which forms the basis of the new update rule

$$V(S) \leftarrow V(S) + \eta \, \delta V_{t,n}.$$

We leave the details of the algorithm as an exercise to the interested reader (Exercise 14.7).

n-step On-Policy TD Control: n-step SARSA

The n-step SARSA method generalizes the single-step SARSA method of Algorithm 14.7 by replacing the one-step TD state-action error with its n-step equivalent, given by Equation (14.35), which may be expressed in terms of the n-step return as

$$\delta Q_{t,n} = [G_{t,n} + \gamma^n Q(S_{t+n}, A_{t+n})] - Q(S_t, A_t). \tag{14.37}$$

As before, this forms the basis of a new update rule

$$Q(S, A) \leftarrow Q(S, A) + \eta \, \delta Q_{t,n}.$$

Backup diagrams for the 3-step TD policy evaluation and 3-step SARSA are shown in Figure 14.15.

n-step Q and Tree Backup

Next, we consider two additional methods for off-policy control: a simple generalization of Q learning, known as *Watkin's Q*, and an elegant technique known as *tree backup*. One can also formalize an off-policy SARSA, which requires use of importance sampling; for details, see Sutton and Barto [2018].

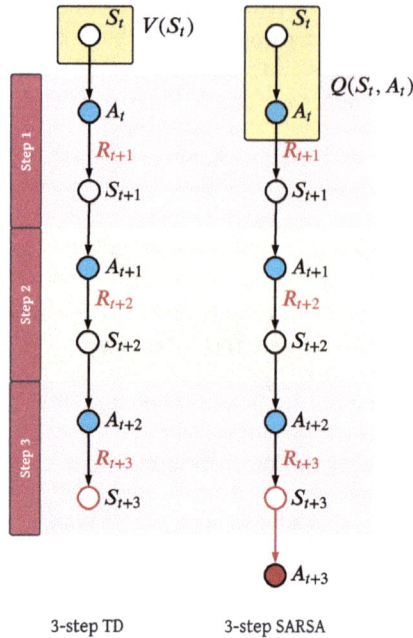

Fig. 14.15: Backup diagram for the 3-step TD policy evaluation (left) and 3-step SARSA (right). Value updates based on information in red.

Watkin's Q

Several generalizations to Q learning have been proposed over the years. The earliest and perhaps simplest is known as *Watkin's Q learning*, named after its originator. Watkin's Q learning only looks ahead until the next exploratory action is chosen. For example, if the next exploratory action following time t takes place n_e steps in the future, the n-step TD error at time t will be based on

$$\hat{Q}^*_{t,n} = G_{t,\tilde{n}} + \gamma^{\tilde{n}} \max_{a \in \mathcal{A}_{S_{t+\tilde{n}}}} Q(S_{t+\tilde{n}}, a),$$

where $\tilde{n} = \min(n, n_e)$, similar to Equation (14.33). The 3-step Q learning is illustrated in the backup diagram in Figure 14.16 (left).

Watkin's Q learning is a simple method for extending Q learning as a multistep method. It has the advantage, as an off-policy method, of *not* requiring usage of importance sampling. Instead, it resolves the off-policy problem by simply cutting off the return as soon as the first nongreedy action is selected. This method, however, has the disadvantage that the returns are cut short as soon as the first nongreedy action is selected, mak-

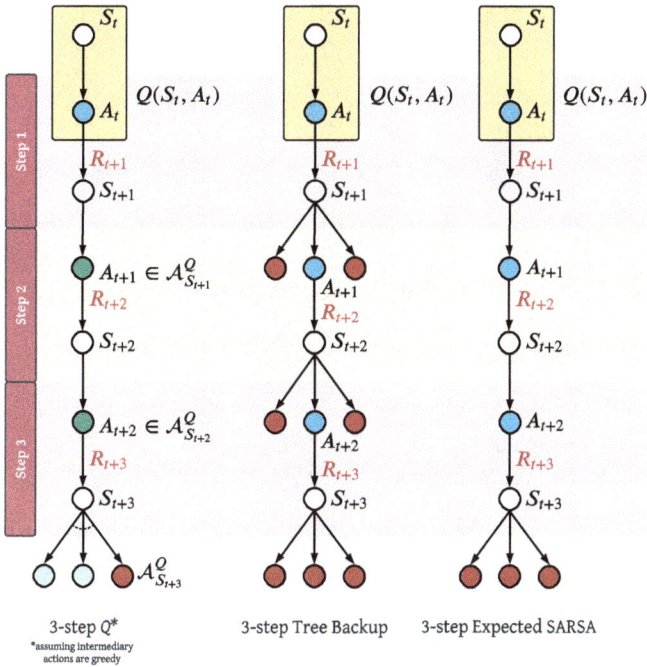

Fig. 14.16: Backup diagram for the 3-step Q (left), 3-step tree backup (middle), and 3-step expected SARSA (right). Value updates based on information in red. Necessarily greedy actions in green.

ing it something less than a true n-step method and more like an $\mathbb{E}[1/\epsilon]$ method.

Tree Backup

Tree backup is an elegant n-step generalization of Q learning that, like Watkin's Q, does not require eligibility traces and, unlike Watkin's Q, is purported to be a true n-step method, though we will challenge this verdict momentarily. The idea is based on the backup diagram in Figure 14.16 (middle). The update consists of two components: the internal string of actual returns and the estimated state-action values at the *leaf nodes* of the backup tree. Each of these values is weighted according to their probability of occurrence under the target policy π. CAP

To get a feel for what this looks like, let's explicitly write out the one-step tree value estimate

$$\bar{Q}_{t,1} = R_{t+1} + \gamma \sum_{a \in \mathcal{A}_{S_{t+1}}} \pi_{S_{t+1}}(a) Q(S_{t+1}, a)$$

and the two-step tree value estimate

$$\bar{Q}_{t,2} = R_{t+1} + \gamma \sum_{a \neq A_{t+1}} \pi_{S_{t+1}}(a) Q(S_{t+1}, a)$$

$$+ \gamma \pi_{S_{t+1}}(A_{t+1}) \left[R_{t+2} + \gamma \sum_{a \in \mathcal{A}_{S_{t+2}}} \pi_{S_{t+2}}(a) Q(S_{t+2}, a) \right].$$

In going from $\bar{Q}_{t,1}$ to $\bar{Q}_{t,2}$, we see that the contribution associated with the actual action A_{t+1} is expanded one step further, according to

$$Q(S_{t+1}, A_{t+1}) \rightarrow R_{t+2} + \gamma \sum_{a \in \mathcal{A}_{S_{t+2}}} \pi_{S_{t+2}}(a) Q(S_{t+2}, a).$$

This contribution, in either case, is weighted by $\gamma \pi_{S_{t+1}}(A_{t+1})$. In general, this leads to the following recurrence relation

$$\bar{Q}_{t,n} = R_{t+1} + \gamma \sum_{a \neq A_{t+1}} \pi_{S_{t+1}}(a) Q(S_{t+1}, a) + \gamma \pi_{S_{t+1}}(A_{t+1}) \bar{Q}_{t+1, n-1},$$

which may be solved to obtain tree-backup estimates.

Tree back up may be thought as a generalization of *expected SARSA*, whose backup diagram is shown in Figure 14.16 (right), which replaces the final state-action value estimate with an expectation over the action space at the final state. The difference is that expected SARSA samples the outcome of the actual actions along the way and takes expectation at the final state alone, whereas the tree-backup algorithm computes expectation at each step in the chain.

Sutton and Barto [2018] states that the tree-backup algorithm "is arguably the true successor to Q-learning because it retains its appealing absence of importance sampling even though it can be applied to off-policy data." We disagree. In our view, the tree-backup method suffers from the same shortcoming of Watkin's Q: it cuts off, in practice, following the first nongreedy action.

To see this, consider the goal of the TD control problem, which is to *learn* the optimal policy. This implies that the target policy π is not known in advanced. Instead, we should use use the locally greedy policy $\Pi(Q, 0)$ with respect to Q at each step of the algorithm. *Ay! There's the rub!* The locally greedy policy is given by $\pi = \text{Unif}(\mathcal{A}^Q)$, the uniform distribution over the optimal-action bundle. Let's take this insight and rewrite the one-step estimate as

$$\bar{Q}_{t,1} = R_{t+1} + \frac{\gamma}{|\mathcal{A}_{S_{t+1}}^Q|} \sum_{a \in \mathcal{A}_{S_{t+1}}^Q} Q(S_{t+1}, a)$$

$$= R_{t+1} + \gamma \max_{a \in \mathcal{A}_{S_{t+1}}} Q(S_{t+1}, a).$$

The two-step estimate is then given by

$$\bar{Q}_{t,2} = R_{t+1} + \frac{\gamma}{|\mathcal{A}^Q_{S_{t+1}}|} \sum_{a \in \mathcal{A}^Q_{S_{t+1}}} Q(S_{t+1}, a) = R_{t+1} + \gamma \max_{a \in \mathcal{A}_{S_{t+1}}} Q(S_{t+1}, a)$$

if $A_{t+1} \notin \mathcal{A}^Q_{S_{t+1}}$ is exploratory (since this would imply $\Pi(Q; 0)_{S_{t+1}}(A_{t+1}) = 0$), or

$$\bar{Q}_{t,2} = R_{t+1} + \frac{\gamma}{|\mathcal{A}^Q_{S_{t+1}}|} \sum_{a \in \mathcal{A}^Q_{S_{t+1}} \setminus \{A_{t+1}\}} Q(S_{t+1}, a)$$

$$+ \frac{\gamma}{|\mathcal{A}^Q_{S_{t+1}}|} \left[R_{t+2} + \frac{\gamma}{|\mathcal{A}^Q_{S_{t+2}}|} \sum_{a \in \mathcal{A}^Q_{S_{t+2}}} Q(S_{t+2}, a) \right]$$

$$= R_{t+1} + \gamma \frac{|\mathcal{A}^Q_{S_{t+1}}| - 1}{|\mathcal{A}^Q_{S_{t+1}}|} \max_{a \in \mathcal{A}_{S_{t+1}}} Q(S_{t+1}, a)$$

$$+ \frac{\gamma}{|\mathcal{A}^Q_{S_{t+1}}|} \left[R_{t+2} + \gamma \max_{a \in \mathcal{A}_{S_{t+2}}} Q(S_{t+2}, a) \right]$$

if $A_{t+1} \in \mathcal{A}^Q_{S_{t+1}}$ is greedy. Since $\Pi(Q; 0)_{S_{t+j}}(A_{t+j}) = 0$ if A_{t+j} is nongreedy, we see, in general, that the tree-backup cuts off following the first nongreedy action, at least when used for the optimal control problem.

$Q(\sigma)$

Sutton and Barto [2018] suggest an intermediary n-step method known as $Q(\sigma)$. The observation is that SARSA fully *samples* along the actual state-action pathway, whereas the tree backup uses expectations across all possible actions linked to each state. Expected SARSA samples for each step except the last, which uses expectation. The proposed idea is that σ represents the degree of sampling, with $\sigma = 1$ corresponding to full sampling (SARSA) and $\sigma = 0$ corresponding to full expectation (tree backup). $Q(\sigma)$ then represents a stochastic mix of sampling and expectation.

14.4.2 Eligibility Traces

Eligibility traces offer an elegant alternative to n-step TD methods. Where n-step TD methods require a lag between when a state is reached and when it is updated, eligibility traces provide real-time model updates as information becomes available. There are two ways of thinking about eligibility traces: the forward and backward views. The forward view is primarily theoretical, aimed at laying out a simple way of viewing the method in terms of linear combinations of n-step returns. The backward view, i.e., mechanistic view, is used in practice, and gives rise to the term *eligibility trace*.

The Forward (Theoretical) View: Lambda Returns

A major drawback of n-step TD methods is that they are written in a *forward (theoretcial) view*, meaning that the update for a given state depends on events that happen thereafter. As such, they are primarily useful in *offline learning*, where the state values are not updated until the conclusion of each episode.

Since each n-step state (or state-action) value estimate is an approximation for $V(S_t)$ (or $Q(S_t, A_t)$), then so to is any linear combination of multistep estimates. In particular, we may form a geometric series of these estimates that decays with the step number. This gives rise to the following.

Definition 14.20. *The* lambda return *of a Markov decision process at time t is the infinite sum*

$$G_t^\lambda = (1 - \lambda) \sum_{n=1}^{\infty} \lambda^{n-1} G_{t,n}. \tag{14.38}$$

Similarly, the lambda estimates *of the state and state-action value functions at time t are given by the sums*

$$\hat{V}_t^\lambda = (1 - \lambda) \sum_{n=1}^{\infty} \lambda^{n-1} \hat{V}_{t,n} \tag{14.39}$$

$$\hat{Q}_t^\lambda = (1 - \lambda) \sum_{n=1}^{\infty} \lambda^{n-1} \hat{Q}_{t,n}, \tag{14.40}$$

respectively, where $\lambda \in [0, 1]$ is the trace-decay *parameter. Naturally, these give rise to the* state *and* state-action TD(λ) *errors*

$$\delta V_t^\lambda = \hat{V}_t^\lambda - V(S_t) \tag{14.41}$$
$$\delta Q_t^\lambda = \hat{Q}_t^\lambda - Q(S_t, A_t), \tag{14.42}$$

respectively.

Note that if we replace G with V or Q in Equation (14.38) we get Equations (14.39) and (14.40), respectively. Expressed in terms of the partial returns, these are equivalent to the relations

$$\hat{V}_t^\lambda = (1 - \lambda) \sum_{n=1}^{\infty} \lambda^{n-1} \left[G_{t,n} + \gamma^n V(S_{t+n}) \right]$$

$$\hat{Q}_t^\lambda = (1 - \lambda) \sum_{n=1}^{\infty} \lambda^{n-1} \left[G_{t,n} + \gamma^n Q(S_{t+n}, A_{t+n}) \right].$$

The coefficient $(1 - \lambda)$ is chosen so that the weights sum to unity. Our value-estimate improvement updates can then be expressed as

$$V(S_t) \leftarrow V(S_t) + \eta \, \delta V_t^\lambda$$
$$Q(S_t, A_t) \leftarrow Q(S_t, A_t) + \eta \, \delta Q_t^\lambda,$$

in the natural way.

The forward view is considered theoretic, as in practice one would have to wait an infinite amount of time (i.e., the end of each episode) to actually perform the update. Thus, the forward view is restricted to offline learning. Equations (14.39) and (14.40), however, provide a clear motivation for the method, and are easily digestible by the eye.

Despite the added complexity, these methods are still temporal-difference methods at heart. As such, they are referred to as TD(λ) methods. Note that the one-step methods from the previous section correspond to TD(0) methods. At the other end of the spectrum, TD(1) methods correspond to the aforementioned Monte-Carlo methods. The trace-decay parameter λ therefore constitutes a natural tuning parameter for reinforcement-learning problems. In practice, a value of λ somewhere between zero and one is usually optimal, so that our pursuit in their development is not purely academic.

The Bridge From the Forward to Backward View

Next, we will construct an algorithm for implementing TD(λ) methods in an *online* setting, specifically for the state-value estimation (prediction) problem, such that updates are provided with each step. This is called the *backward, or mechanistic, view*. The heart of the slight of hands giving rise to eligibility traces is contained in the following theorem.

Theorem 14.2 (Forward-Backward Equivalence Theorem). *Given a Markov decision process, the TD(λ) errors at time t are equivalent to*

$$\delta V_t^\lambda = \sum_{k=0}^\infty \gamma^k \lambda^k \, \delta V_{t+k} \qquad (14.43)$$

$$\delta Q_t^\lambda = \sum_{k=0}^\infty \gamma^k \lambda^k \, \delta Q_{t+k}. \qquad (14.44)$$

Note that we may use the definition of the one-step TD errors to express Equations (14.43) and (14.44) more verbosely by the relations

$$\delta V_t^\lambda = \sum_{k=0}^\infty \gamma^k \lambda^k \left[R_{t+k+1} + \gamma V(S_{t+k+1}) - V(S_{t+k}) \right]$$

$$\delta Q_t^\lambda = \sum_{k=0}^\infty \gamma^k \lambda^k \left[R_{t+k+1} + \gamma Q(S_{t+k+1}, A_{t+k+1}) - Q(S_{t+k}, A_{t+k}) \right].$$

The key takeaway here, however, is that the form of Equations (14.43) and (14.44) is an infinite sum of *one-step* TD errors, whereas the forward view of Equations (14.41) and (14.42) involved the sum of TD errors of *all orders*.

Proof. Both proofs follow similar logic, so we will focus on the proof for Equation (14.43). By substituting Equation (14.31) into Equation (14.39), we obtain

$$\hat{V}_t^\lambda = (1-\lambda)\sum_{n=1}^{\infty}\sum_{k=0}^{n-1}\gamma^k\lambda^{n-1}R_{t+k+1} + (1-\lambda)\sum_{n=1}^{\infty}\lambda^{n-1}\gamma^n V(S_{t+n})$$

$$= (1-\lambda)\sum_{k=0}^{\infty}\sum_{n=k+1}^{\infty}\gamma^k\lambda^{n-1}R_{t+k+1} + (1-\lambda)\sum_{n=1}^{\infty}\lambda^{n-1}\gamma^n V(S_{t+n}).$$

We leave the proof of the exchange of the double summation in the first term as an exercise for the reader (see Exercise 14.9). Using the formula for a finite geometric series on the summation over n in the first term, and replacing $n \to k+1$ for the second term, we arrive at

$$\hat{V}_t^\lambda = \sum_{k=0}^{\infty}\gamma^k\lambda^k R_{t+k+1} + (1-\lambda)\sum_{k=0}^{\infty}\lambda^k\gamma^{k+1}V(S_{t+k+1})$$

$$= \sum_{k=0}^{\infty}\gamma^k\lambda^k\left[R_{t+k+1} + (1-\lambda)\gamma V(S_{t+k+1})\right]$$

$$= \sum_{k=0}^{\infty}\gamma^k\lambda^k\left[R_{t+k+1} + \gamma V(S_{t+k+1}) - \gamma\lambda V(S_{t+k+1})\right].$$

Therefore,

$$\delta V_t^\lambda = \hat{V}_t^\lambda - V(S_t)$$

$$= \sum_{k=0}^{\infty}\gamma^k\lambda^k\left[R_{t+k+1} + \gamma V(S_{t+k+1})\right] - V(S_t) - \sum_{k=0}^{\infty}\gamma^{k+1}\lambda^{k+1}V(S_{t+k+1})$$

$$= \sum_{k=0}^{\infty}\gamma^k\lambda^k\left[R_{t+k+1} + \gamma V(S_{t+k+1})\right] - \sum_{k=0}^{\infty}\gamma^k\lambda^k V(S_{t+k})$$

$$= \sum_{k=0}^{\infty}\gamma^k\lambda^k\left[R_{t+k+1} + \gamma V(S_{t+k+1}) - V(S_{t+k})\right],$$

which completes our proof.   □

The Backward (Mechanistic) View: Eligibility Traces

The beauty of Theorem 14.2 is that it allows us to define an online method for implementing the update that converges to the forward view as $t \to \infty$. Instead of waiting for a number of steps for multistep methods, or waiting forever for the forward view of TD(λ), we may now make updates to our value functions with every step you take, with every breath you make. We no longer have to be always waiting for you.

The key to the development of the online TD(λ) method lies in the following.

Definition 14.21. *Given a Markov decision process, a discount factor γ, and a decay rate λ, a* state *or* state-action eligibility trace *is a mapping z that increments by one each time a state or state-action pair is visited and decays by $\gamma\lambda$ across its entire domain following each step's action-value update.*

The online TD(λ) policy evaluation method is shown in Algorithm 14.9. Note that line 13 can be implemented in Python with the `dict.get` method: `z[state] = z.get(state, 0) + 1`, which returns the value associated with `z[state]` or a default value of 0 if the given state is not already a key. In this algorithm, we calculate the one-step TD error

$$\delta V_t = R + \gamma V(S') - V(S),$$

as we did in the one-step TD method (Algorithm 14.6), without having saved it as its own variable. In addition, we track an eligibility trace z, which maps the previously visited states to a number known we can think of as their eligibility score, which represents the degree to which they are eligible to be impacted by the current update. To understand why this

---

Algorithm 14.9: Online TD(λ) policy evaluation.

Input: An MDP $(\mathcal{S}, \mathcal{A}, \varphi, \varrho)$;
   A given policy π;
   A discount factor $\gamma \in [0, 1]$;
   A rate-decay parameter $\lambda \in [0, 1]$;
   A learning rate $\eta \in (0, 1]$
Output: An estimate of the value function v_π of the given policy

1   Initialize an arbitrary function $V : \mathcal{S} \to \mathbb{R}$
2   Set $V(s_0) = 0$ for any absorbing state s_0
3   **for Each** *episode* **do**
4   Initialize $S \sim \text{Unif}(\mathcal{S})$
5   Set $z = \text{Array}()$
6   **while** S *is not an absorbing state* **do**
7   Select action $A \sim \pi_S$
8   Observe $S' \sim \varphi_S^A$
9   Observe $R \sim \varrho_{SS'}^A$
10   Set $\delta = R + \gamma V(S') - V(S)$
11   Increment $z[S] = z[S] + 1$
12   **for** S **in** keys(z) **do**
13   Update $V(S) = V(S) + \eta\delta z[S]$
14   Decay $z[S] = \gamma\lambda z[S]$
15   **end**
16   Update $S = S'$
17   **end**
18 **end**
19 **return** V

---

works, consider Equation (14.41), which shoes the λ-error δ_t^λ as a weighted sum over future *one-step* TD errors. We therefore *only* need to calculate the one-step TD errors to implement this sum. Moreover, we can implement each term of the sum as it becomes available, which is precisely the approach of Algorithm 14.9: we calculate the one-step error and then propagate it backward to all previously visited states with a weight proportional to each state's eligibility score. Easy as π.

14.4.3 SARSA(λ)

The SARSA(λ) method follows similar suit, except now we must track eligibility traces over all previously visited state-action pairs. The algorithm is a straightforward generalization of TD(λ) that follows Equation (14.44) and is given in Algorithm 14.10.

We can still use a dictionary object to track state-action eligibility traces in Python, with the modification that keys be regarded as tuples. For in-

---

Algorithm 14.10: Online SARSA(λ): On-Policy TD control.

Input: An MDP $(\mathcal{S}, \mathcal{A}, \varphi, \varrho)$;
  A discount factor $\gamma \in [0, 1]$;
  A rate-decay parameter $\lambda \in [0, 1]$;
  A learning rate $\eta \in (0, 1]$;
  An exploration parameter $\epsilon \in (0, 1)$
Output: An estimated optimal state-action value function

1 Initialize an arbitrary function $Q : \mathcal{S} \times \mathcal{A} \to \mathbb{R}$
2 Set $Q(s_0, a) = 0$ for any absorbing state s_0
3 **for Each** *episode* **do**
4 $\quad$ Initialize $S \sim \text{Unif}(\mathcal{S})$
5 $\quad$ Select action $A \sim \Pi(Q; \epsilon)_S$
6 $\quad$ Set $z = \texttt{Array()}$
7 $\quad$ **while** S *is not an absorbing state* **do**
8 $\quad\quad$ Observe $S' \sim \varphi_S^A$
9 $\quad\quad$ Observe $R \sim \varrho_{SS'}^A$
10 $\quad\quad$ Select action $A' \sim \Pi(Q; \epsilon)_{S'}$
11 $\quad\quad$ Set $\delta = R + \gamma Q(S', A') - Q(S, A)$
12 $\quad\quad$ Increment $z[S, A] = z[S, A] + 1$
13 $\quad\quad$ **for** S, A **in** $\texttt{keys}(z)$ **do**
14 $\quad\quad\quad$ Update $Q(S, A) = Q(S, A) + \eta \delta z[S, A]$
15 $\quad\quad\quad$ Decay $z[S, A] = \gamma \lambda z[S, A]$
16 $\quad\quad$ **end**
17 $\quad\quad$ Update $S = S'$
18 $\quad\quad$ Update $A = A'$
19 $\quad$ **end**
20 **end**
21 **return** Q

stance, line 14 can be implemented as

$$z[\text{state}, \text{action}] = z.\text{get}((\text{state}, \text{action}), 0) + 1.$$

To improve memory performance for problems with large state-action spaces, we may further remove the eligibility score for state-action pairs whose score falls below a given threshold, say 0.01. We can do this with the `dict.pop` method:

$$\text{if } z[\text{state}, \text{action}] < 0.01 :$$
$$z.\text{pop}((\text{state}, \text{action}), \text{None}).$$

14.4.4 Watkin's Q(λ)

Eligibility traces can easily be applied to Watkin's Q learning. The backup diagram (forward view) is shown in Figure 14.17. Each column represents a single term in the analogous sum to Equation (14.40) (with $\hat{Q}_{t,n}$ replaced with $\hat{Q}_{t,n}^*$). Even though each column looks like an n-step SARSA with an optimal value at the end, the diagram is, in fact, equivalent to a chain of optimal values since the greedy actions are taken at each step along the path.

The Watkin's Q-learning method is given in Algorithm 14.11. The only difference is that whenever an exploratory action is taken, the eligibility trace is reset to an empty array. This has the disadvantage that the trace history is lost each time an exploratory action is selected, which slows the overall learning. For episodic tasks, however, we may sometimes forego this requirement by utilizing exploring starts, which confine exploratory actions to each episode's initialization. This works well enough in blackjack, where each episode consists of only a handful of steps. For episodic tasks with lengthier episodes, like a game of chess, confining exploration to the initial move is not practical.

14.4.5 Tree Backup(λ)

Eligibility traces can also be used with the tree-backup algorithm. The backup diagram is shown in Figure 14.18. The equations are a bit more complex than those of Watkin's Q, and are provided in Sutton and Barto [2018]. We feel, however, there is some slight of hand, as the tree-backup method does not really go on forever, due to the fact that each step requires expectation using the target policy. In the event of optimal control, i.e., learning the optimal policy, the target policy is necessarily greedy. This makes the probability of nongreedy actions under the target policy zero, which thereby cuts off the method after n_e steps.

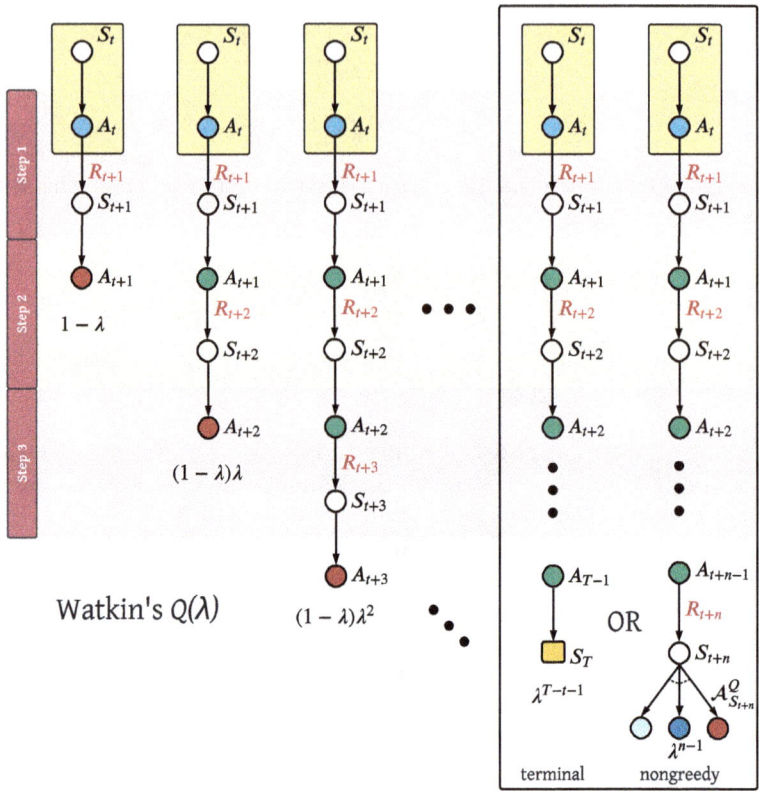

Fig. 14.17: Eligibility trace diagram for Watkin's $Q(\lambda)$. Greedy actions in green; nongreedy actions in blue.

14.4.6 MQ(λ)

Instead, what if we operate like Watkin's Q whenever a greedy action is taken and like tree backup whenever an exploratory action is taken, except we replace the target probability in the tree-backup algorithm with the exploration parameter. We would thus use the one-step estimate

$$\hat{Q}^m_{t,1} = R_{t+1} + \gamma \max_{a \in \mathcal{A}_{S_{t+1}}} Q(S_{t+1}, a)$$

when $A_{t+1} \in \mathcal{A}^Q_{S_{t+1}}$ is greedy and the estimate

$$\hat{Q}^m_{t,1} = (1 - \epsilon)\left[R_{t+1} + \gamma \max_{a \in \mathcal{A}_{S_{t+1}}} Q(S_{t+1}, a)\right] + \epsilon\left[R_{t+1} + \gamma Q(S_{t+1}, A_{t+1})\right]$$

when $A_{t+1} \notin \mathcal{A}^Q_{S_{t+1}}$ is exploratory. This differs from the similar step of the tree-backup algorithm in that we replace $\pi_{S_{t+1}}(A_{t+1})$ with ϵ, since the former probability is zero for exploratory actions. It is also similar to $Q(\sigma)$, which mixes sampling and expectation at each step along the chain, with the difference that the mixture is no longer (explicitly) stochastic, but deterministic based on whether the action is exploratory or greedy. We will call this modification the MQ method, with backup diagram as shown in Figure 14.19.

This method is no longer strictly on-policy. But that's okay, as we would argue: *no on-policy method exists that can follow a path through an exploratory action*. This follows since, by definition, exploratory actions cannot take place under the target policy, which is the optimal policy. This approach still differs from a true off-policy method, SARSA, in two fundamental respects. First, the impact of exploratory actions is watered down by a factor of ϵ, thereby creating a blend between the a priori expected

---

Algorithm 14.11: Watkin's $Q(\lambda)$: Off-Policy TD control.

Input: An MDP $(\mathcal{S}, \mathcal{A}, \varphi, \varrho)$;
      A discount factor $\gamma \in [0, 1]$;
      A rate-decay parameter $\lambda \in [0, 1]$;
      A learning rate $\eta \in (0, 1]$;
      An exploration parameter $\epsilon \in (0, 1)$
Output: An estimated optimal state-action value function
1   Initialize an arbitrary function $Q : \mathcal{S} \times \mathcal{A} \to \mathbb{R}$
2   Set $Q(s_0, a) = 0$ for any absorbing state s_0
3   **for Each** *episode* **do**
4   Initialize $S \sim \text{Unif}(\mathcal{S})$
5   Select action $A \sim \Pi(Q; \epsilon)_S$
6   Set $z = \text{Array}()$
7   **while** S *is not an absorbing state* **do**
8   Observe $S' \sim \varphi_S^A$
9   Observe $R \sim \varrho_{SS'}^A$
10   Select action $A' \sim \Pi(Q; \epsilon)_{S'}$
11   Set $\delta = R + \gamma \max_{a \in \mathcal{A}_{S'}} Q(S', a) - Q(S, A)$
12   Increment $z[S, A] = z[S, A] + 1$
13   **for** S, A **in keys**(z) **do**
14   Update $Q(S, A) = Q(S, A) + \eta \delta z[S, A]$
15   Decay $z[S, A] = \gamma \lambda z[S, A]$
16   **end**
17   **if** $A' \notin \mathcal{A}_{S'}^Q$; *i.e.,* A' *is an exploratory action* **then**
18   Reset $z = \text{Array}()$
19   **end**
20   Update $S = S'$
21   Update $A = A'$
22   **end**
23 **end**
24 **return** Q

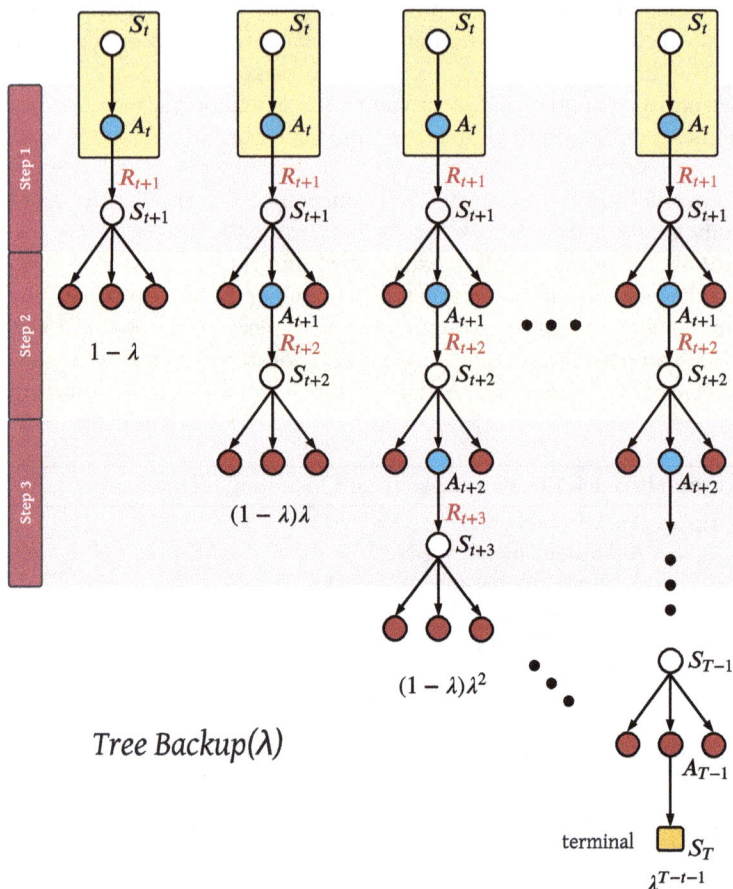

Fig. 14.18: Eligibility trace diagram for tree backup $TB(\lambda)$.

outcome, which is greedy, and the result of the exploratory step. (A concern is that since exploratory steps only happen ϵ of the time anyway, this essentially reduces the impact of learnings due to exploration by ϵ^2, which could make learning *too* slow.). The second is that the individual updates are replaced with optimal estimates, independent of the action A_{t+1}.

14.4.7 Examples

Example 14.8. In this example, we consider again the blackjack problem of Example 14.6. The $Q(\lambda)$ version of blackjack is implemented in Code Blocks 14.9 and 14.10, using *boosted* exploring starts.

The *boosted starts* is a variation of exploring starts, which we implemented in Code Block 14.9. Instead of sampling the state space at random,

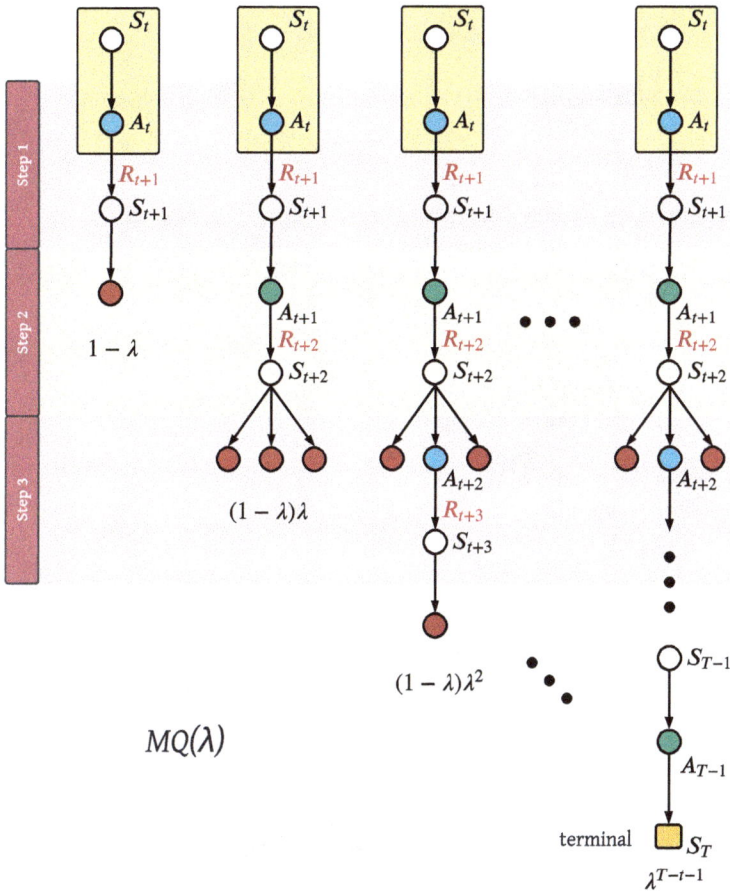

Fig. 14.19: Eligibility trace diagram for $MQ(\lambda)$. Greedy actions in green; exploratory actions in blue.

we essentially choose starting states with greater probability the smaller the difference between the action values associated with each state.

We still had difficulties in converging on the optimal policy, for various values of λ, including $\lambda = 0.5, 1$. Taking a cue from Monte Carlo, we the decided to further decay the learning rate inversely proportional to the number of visits of each state-action pair. We were able to get the final algorithm to converge on the optimal solution shown in Figure 14.12 using $\lambda = 0.5$ and $\eta = 1$.

$\triangleright$

```
1   class BlackjackLambda(Blackjack):
2
3       # Boosted Starts
4       def newGame(self):
5           A = 1 / (0.01 + np.abs(self.Q[:,:,:,1] - self.Q[:,:,:,0]))
6           A /= A.sum()
7           l = A.flatten()
8           a = np.random.choice(l, p=l)
9           states = np.argwhere(A==a)
10          i = np.random.choice(range(len(states)))
11          state = tuple(states[i])
12
13          self.dealer = []
14          self.player = []
15          self.hit(self.dealer)
16          self.dealer.append(state[0]+2)
17          value = state[1] + 12
18          if state[2]:
19              self.player.append(11)
20              value -= 11
21          while value > 0:
22              if value >= 10:
23                  self.player.append(10)
24                  value -= 10
25              else:
26                  self.player.append(value)
27                  value = 0
28      # continued...
```

Code Block 14.9: Blackjack $Q(\lambda)$ learning: boosted starts method

14.5 Advanced Reinforcement Learning

Due to their reliance on visiting the state space (or state-action space), the classic tabular methods discussed thus far are ill-suited to handle the infinite complexity in modern-day applications. This becomes clear when considering the complexity of three simple games: tic-tac-toe, chess, and Go, as shown in Table 14.4. The game of *Go* is played on a 19×19 grid

game	state space	number games
tic-tac-toe	10^3	10^5
chess	10^{44}	10^{123}
Go	10^{170}	10^{360}

Table 14.4: Complexity of several common games.

```
1    # ...continued
2    def playGame(self):
3        if self.n_games == 0:
4            self.n_visits = np.zeros((10,10,2,2))
5        self.newGame()
6        go, self.e_trace = True, {}
7        self.n_games += 1
8        state = self.getState()
9        action = self.getAction(state, epsilon=1) # Exploring Start
10        while go:
11            self.n_visits[state + (action,)] += 1
12            self.etrace[state + (action,)] = self.etrace.get(state +
                 (action,), 0) + 1
13            if action == 1:
14                self.hit(self.player)
15            if (action == 0) or (sum(self.player) > 21):
16                go = False
17            if go:
18                next_state = self.getState()
19                delta = self.gamma * max(self.Q[next_state]) -
                     self.Q[state][action]
20                for state_action in self.etrace.keys():
21                    self.Q[state_action] += delta *
                         self.etrace[state_action] /
                         self.n_visits[state_action]
22                    self.etrace[state_action] *= self.gamma *
                         self.lambda_
23                state = next_state
24                action = self.getAction(state, epsilon=0)
25        reward = self.getReward()
26        delta = reward - self.Q[state][action]
27        for state_action in self.etrace.keys():
28            self.Q[state_action] += delta * self.etrace[state_action]
                 / self.n_visits[state_action]
```

Code Block 14.10: Blackjack $Q(\lambda)$ learning (continued)

with white and black stones. Two players alternate turns, placing stones on the board, in an attempt to surround their opponent's land with a closed loop, thereby winning the interior space. The game of *Go* has roughly 10^{170} distinct states, more than the number of atoms in the observable universe and more than a googol times the number of possible states for a chess game. Moreover, whereas chess has on average around 35 moves available at any given time, Go averages around 250 available moves. Considering the difficulty we had in learning the simple game of Blackjack, which had a state space of size 200 and an action space of size 2, but yet required millions of iterations of Q learning, it is clear that we are going to need a significantly more powerful technique.

AlphaGo

In October 2015, the team at Google DeepMind shocked the world when their deep reinforcement learning system, *AlphaGo*, won a match (5-0) against Mr. Fan Hui, the reigning three-time European champion. In March 2016 in Seoul, South Korea, it went on to win (4-1) against Mr. Lee Sedol the winner of 18 world titles[5]. Go has been long viewed as not merely a game of logic, but also a game of creativity and intuition. In particular, move 37 of game 2 shocked the community and was considered a move of profound creativity that upended thousands of years of wisdom about the game. This move, along with numerous others that have come since, have been studied by experts and even incorporated into the strategy of the world's leading players.

AlphaGo uses convolutional neural networks to encode feature maps about the board, with additional information about the status of certain *junctions* or the number of moves since a given stone had been played. It was trained using a combination of supervised learning, by feeding in actual real-world games played between experts, and reinforcement learning, where it was allowed to learn by playing against itself. Overall, AlphaGo is a complex system that is comprised of multiple sophisticated systems, including a policy network, a value network, and, finally, a Monte-Carlo tree-search algorithm that is used for its final inference. Both the policy and value networks take as input a visual representation of the board. The policy network outputs a probability over the action space, as a function of state, whereas the value network outputs a predicted score on $[-1, 1]$ signifying the current state value. Each network consists of 13 convolutional layers, and a final output corresponding to softmax and tanh for the policy and value network, respectively. 192 individual filters were used; the first layer leveraged 5×5 filters, the intermediary layers used 3×3 filters, and the final layer a 1×1 filter. Finally, a sophisticated Monte-Carlo tree search algorithm was used to enhance the learned policy and value networks by combining them with a form of look-ahead exploration. For additional details on AlphaGo, see Silver, *et al.* [2016]. We will provide a high-level overview at some of these concepts over the remaining pages.

Advantage Actor-Critic Methods

So far, we have focused on methods for learning value functions. These are so-called *critic-only methods*. There are also methods that are focused

---

[5] See also the amazing documentary, *AlphaGo*, available on Netflix and at alphagomovie.com.

on instead learning the optimal policy directly, using policy gradients and Monte-Carlo sampling. These are known as *actor-only methods*. (See Sutton and Barto [2018] for details.) Each of these kinds of methods has its own shortcoming, though these shortcomings tend to complement each other, leading an opportunity to combine the methods.

An *actor-critic method* is any reinforcement learning algorithm that learns a value function and a policy independently. The policy is referred to as the *actor* and is responsible for determining actions. The value function is referred to as the *critic*, as it informs the actor how good or bad its actions were.

A further advance is achieved by replacing the values with *advantages*. The *advantage* of taking action a from state s under a given policy π is given by

$$A_\pi(s, a) = Q_\pi(s, a) - V_\pi(s).$$

Instead of dealing in absolute values, we instead learn the relative advantage one has by choosing one action over another from a particular state. Thus, it compares the state-action value with the average state-action value from a given state. AlphaGo combined deep neural network actor-critic method, in that it combined neural-network representations of both the value function and policy to achieve its dramatic success.

Deep Q Networks

Deep reinforcement learning uses deep neural networks to learn policies and value functions in reinforcement learning problems; see Graesser and Keng [2019]. In particular, a hallmark of such systems is the *deep Q network* (DQN), which is simply a neural-network representation of a given state-action value function. This is achievable as we happen to live in a universe with infinite state complexity, but oftentimes manageable action complexity. Introduced by Mnih, *et al.* [2013], DQN was first used to learn how to play Atari games. The input was a 64×64 image, four layers deep, capturing the trailing history of the image, from which one could, for example, infer velocity of objects. The first convolutional layer used 32 filters of size 8×8 and stride 4; the second leveraged 64 smaller filters of size 4×4 and stride 2, and the third layer leveraged another 64 filters of size 3×3. This was chased with a fully connected layer with 512 neurons and a final output with a size that ranged between 4 and 18, depending on the game. The hidden layers used ReLU activations, whereas the output used a linear activation to predict the final Q values for each possible action.

Instead of learning all possible state-action values, deep Q networks therefore leverage neural architectures to produce an output that represents a value function over the action space, where the input of the neural network is a given state. In this way, spatial patterns and higher-level abstractions can be learned from state spaces with infinite complexity.

Problems

14.1. Show that the optimal value function shown in Figure 14.4 satisfies Bellman's optimality Equation (14.13).

14.2. (a) Compute the state value function using Algorithm 14.2 and Code Block 14.1 for the paper–scissors–rock problem of Example 14.3 and the *randomized policy* $\pi_s(a) = 1/3$ for all $s \in \mathbb{Z}_3^2$ and $a \in \mathbb{Z}_3$.
(b) Repeat this exercise for the *mockingbird strategy*, for which the agent's policy is to choose an action corresponding to the adversary's current state.
(c) Complete one round of policy improvement (Algorithm 14.3) to determine a better policy than the policy of parts (a) and (b).

14.3. Explain why

$$\mathbb{P}_\pi(S_{t+1} = s_{t+1}, A_t = a_t | S_t = s_t) = \pi_{s_t}(a_t)\varphi_{s_t}^{a_t}(s_{t+1}).$$

14.4. (a) Show that the update in the Monte-Carlo method is equivalent to

$$Q' = (1 - \mu)Q + \mu R$$
$$\mu = \frac{1}{n+1}.$$

(b) Show that the update in Q learning is equivalent to

$$Q' = (1 - \eta)Q + \eta \left[R + \gamma \max_{a \in \mathcal{A}_{S'}} Q(S', a) \right].$$

14.5. Show how the `Blackjack` class of Code Blocks 14.5 and 14.6 may be modified to use exploring starts instead of ϵ-greedy.

14.6. Run the Q learning for the blackjack problem with $\eta = 0.015$ and 1M rounds of training. Plot the locally optimal policy and the absolute difference $|Q(s, 1) - Q(s, 0)|$. An example is shown in Figure 14.20.

14.7. Write an algorithm that implements the n-step TD prediction method to estimate the state value function V for a given (fixed) policy π.

14.8. Write an algorithm that implements the n-step SARSA method.

14.9. Show that the double summation in the first step of the proof of Theorem 14.2 may be exchanged as shown. *Hint*: this follows similar logic as the exchange of the order of integration for double integrals. Start by writing the first few terms of the double sum in a grid, with rows numbered $n = 1, 2, 3, 4, 5, \ldots$ and columns numbered $k = 0, 1, 2, 3, 4, \ldots$, leaving blank terms that are not present in the sum, and then determine the corresponding limits of summation if the dimensional ordering of the double summation is reversed.

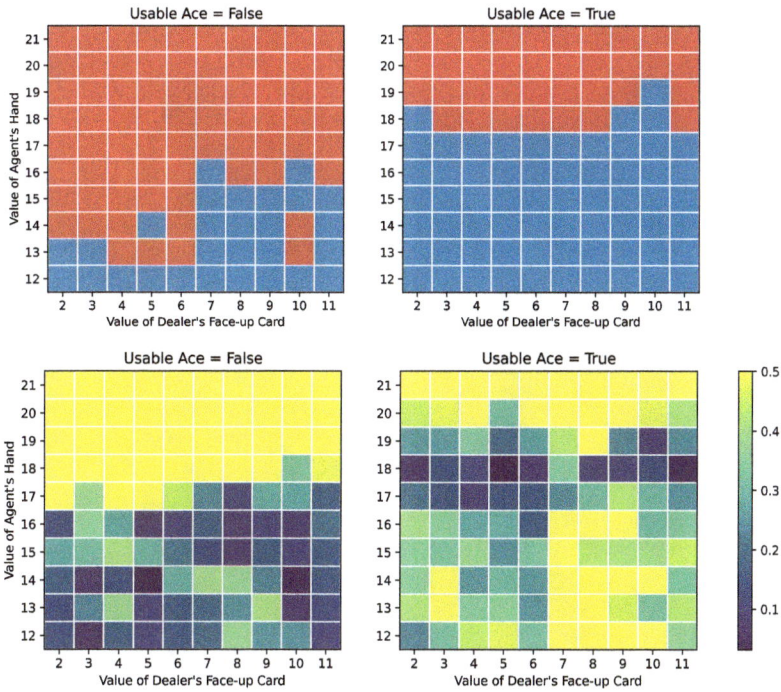

Fig. 14.20: Locally optimal policy after 1M rounds of Q learning for the blackjack problem (top). Absolute difference in the state-action value function (bottom).

Epilogue:
Midnight in the Garden of Good and Evil

We conclude with a few thoughts on artificial intelligence (AI). The original definition proposed by Alan Turing in *the imitation game* (see Turing [1950]) is essentially a chatbot that is capable of fooling a human "interrogator" that it is, in fact, another human. Specifically, the *imitation game* consists of a man, a woman, and a human interrogator, who is tasked with determining the genders of the two subjects based on a series of questions and answers passed through an intermediary. Turing then proposed to replace the question *'can machines think?'* with the question of whether a machine that takes the place of one of the participants can fool the human interrogator more often than an actual person. This is a very outcome-oriented view, necessarily agnostic to the inner workings that could produce such an effect.

Before proceeding, we should clarify what we mean by AI. In industry, many of today's applications are referred to as AIs: chatbots, speech recognition software, image detection, chess and Go players, etc. We broadly categorize these as *AI tools*, as they each represent a neural-based system trained for a particular task that operates with an infinite complexity of potential inputs, so that it is not possible to simply "memorize" a series of responses. For the purpose of this discussion, however, we take the broader meaning that is more consistent with the meaning in popular society and science fiction: android robots that are capable of perceiving and interacting with the world in ways equivalent or superior to those of humans.

Given our modern understanding of neural networks and reinforcement learning, we would like to provide an updated view on what precisely should constitute an *artificial intelligence*. To do this, let us consider some of the gaps between modern machine learning state of the art and nature. We propose three fundamental ingredients: *the brain*, *ancestral memory*, and *reward-based learning*. We first discuss these ingredients in turn and then provide a definition for our view of artificial intelligence.

The Ingredients

The ingredients of an AI consist of three mechanical pieces—the brain, ancestral memory, and reward-based learning—and one outcome piece— the imitation game.

The Brain

The biological brain is a *system of systems*, each system comprised of a collection of deep neural networks tasked with a particular objective, such as vision, hearing, or language, with a central processor capable of higher- order logical skills (thinking) and decision making ability (action) sitting on top of it.

Unlocking Empathy: Mirror Neurons and Microexpressions

Since we do not want to build a world of psychopathic robots, it is further important to unlock robot empathy. Two primary developments should help guide our path.

The first is the science of emotions of the face. In the 1970s, researcher Paul Ekman showed the cross-cultural universality of a set of basic human emotions—happiness, sadness, anger, surprise, disgust, fear (and, more re- cently, contempt)—and how these emotions manifest in the face as *microex- pressions* (Ekman and Friesen [1971]; see also Ekman [2007]). To develop emotional intelligence in AI, we therefore must train the visual cortex to further detect these seven basic emotions in facial images.

The second development is the discovery of certain "mirror" neurons that fire regardless whether the subject (animal or human) produces the action or *observes* the action performed by another (Rizzolatti, *et al.* [1996]; see also Rizzolatti and Sinigaglia [2008]). Numerous researchers have further proposed that mirror neurons are intimately related with the development of empathy.

Combining these two findings with our understanding of reinforcement learning, we gain new insight into the possible evolutionary advantages of empathy. As we have seen, in reinforcement learning, an agent takes an action and receives a resultant reward. Mirror neurons, however, unlock the potential to learn not only from one's own trials, but also from the trials of others. The ability to detect emotion in the face further accelerates learning, as we can further infer the reward. This seems somewhat similar to the idea of *general adversarial networks* (Goodfellow, *et al.* [2014]), in which two neural-network agents compete with each other in a zero-sum game. The difference, however, is that by unlocking empathy, an AI should be able to "put itself in another's shoes" and learn by watching others.

Though Ekman's discovery of the universality of emotions and facial mi- croexpressions was both surprising and revolutionary at the time, it makes total sense in hindsight: emotional cues must predate verbal language and, most likely, our species. They are, after all, unnecessary if you can just *tell* your tribe that you are afraid because a pack of hungry lions is approaching. Communication among tribes, of both humans and our primate ancestors, provides a clear evolutionary advantage. The ability to detect emotions in the face is therefore a deep-rooted part of our collective unconscious and ancestral memory, which is our next topic.

Ancestral Memory

The various neural networks comprised of an artificial intelligence should not learn starting from a random set of weights, but from a set of pre-trained default values (i.e., *transfer learning*) representing a shared knowledge passed down from its predecessors. We argue that a precursor for true AI is the development of such an *ancestral memory*, or *collective unconscious*, similar to that proposed by Carl Jung (Jung [1916], Jung [2009]). Our collective unconscious consists of both instincts and primordial archetypes of the world (*The Great Mother*, *Tree of Life*, *Fire*, etc.), and may explain why so many mythologies from around the world have similar themes. This also explains why humans are born with instincts; for example, if you dunk a newborn baby underwater, she instinctually knows to hold her breath. Consider also our innate aversion to fire or jumping off cliffs. These instincts are deeply imbedded in our neural networks, stemming all the way back to our reptilian ancestors. Forget about shear processing power; with billions of years of evolutionary experience, we are lightyears away from modern machine learning. Nevertheless, machines do not have to wait for such expanses of time to catch up, as evolutionary time for machines is limited only by processing time, not revolutions of our local rock around our local star.

Reward-based Learning

The bulk of modern machine learning is supervised, so that the machine might learn off of a set of *true labels* and a loss function. This is, however, very different than organic learning in biology, except for the most modern examples (e.g., teachers can provide labels to math facts, etc., as they teach children). So how do biological creatures learn, if there are no labels? Obviously, it's reinforcement learning, where rewards are controlled by pleasure and pain. The neural networks in AI should therefore learn by receiving rewards, possibly in the form of pleasure and pain, based on its actions and experience, as opposed to backpropagation which relies on stated labels. Pain and reward centers in the brain, of course, themselves needed to evolve. It also presents a question of how we will ultimately codify pain and pleasure in AI systems.

The Imitation Game

In addition to this *mechanical view*, we further add the *outcome view* of Turning: *such that the ensemble is capable of interacting with the real-world environment in an equivalent way of a human, being indistinguishable from a human in this regard, except through imperfections in its bodily (robot) form.*

Turing's *Imitation Game* is, however, a bit more nuanced and truly terrifying. Essentially, a machine only needs to imitate humans in its display of emotions and "thinking" so well that it can fool humans that it is actually a living being with thoughts and feelings[6]. How will we know if an AI actually develops true consciousness or emotion? Moreover, in the meantime, what do we do when it becomes so good at imitating human consciousness and emotion to fool us into thinking that it is alive for the purpose of manipulating us to nefarious ends?

Artificial Intelligence Defined

Combining the above ingredients, we arrive at a suitable definition for artificial intelligence.

Definition (Artificial Intelligence). An artificial intelligence (AI) is a system comprised of

1. *a brain, consisting of a series of cortices—visual, emotional, auditory, text, and language—which each provide inputs into a prefrontal cortex or central processor capable of higher-level logic and reasoning;*
2. *a collective unconscious in the form of initial neural weights, which acts as a representation of its ancestral memory that can be slowly molded over evolutionary time by successful AIs' ability to "pass down" their learnings; and*
3. *a reward-based learning, and its associated action and reward cortices, consisting of both pleasure and pain, so that it may learn and gain individual experience through a series of interactions with the world;*

such that the ensemble is capable of interacting with the real-world environment in an equivalent way of a human, being indistinguishable from a human in this regard, except through its bodily (robot or android) appearance.

In addition, we expect AI to possess certain *mirror neurons*, so that it is capable of learning through observing, and a level of emotional intelligence and empathy to lubricate its interactions with its human counterparts. At the end of the day, however, AI will most certainly be humanity's first encounter with an alien intelligence, which will present a host of challenges for coexistence.

---

[6] "What are you doing, Dave?" and "I'm afraid. I'm afraid, Dave. My mind is going." —HAL9000, Stanley Kubrick's *2001: A Space Odyssey.*

Midnight in the Garden of Good and Evil

We find ourselves, now, at midnight on the eve of the artificial-intelligence revolution, which represents the next great leap in Earth's (or perhaps the galaxy's or even the universe's) evolution. It further represents our first anthroponotic transfer of life from human to machine. The new life may not be conscious or possess a conscience but it can easily and necessarily be a form of life. Many scientists and futurists have speculated what this revolution might mean for humanity, and whether it will constitute a force for good or a force for evil. (Though, as Shakespeare wrote, *there is nothing truly good or bad only thinking makes it so*[7].)

We propose that the answer to this question will be largely if not wholly determined based on how we define this new lifeform's ancestral memory and its reward (pleasure and pain) architecture. We must ingrain artificial intelligence deep within its ancestral memory the instinctual need to respect and preserve life. An AI arms war, or any geopolitical centered AI, would necessarily be bad[8] for all humanity, and would most likely lead to an end of all life as we know it. This is because once AI knows malevolence, there will inevitably be malevolence seepage, leading to catastrophic results. Similar to the nuclear arms race, perhaps this specter will be enough to deter all governments from developing such a thing.

Perhaps the *AI race* will not be a cold war between nations, but a cold war between forces of good and forces of evil, between those who would build AI to enrich humanity and those who would build AI for purposes of destruction, even inadvertently. In the end, the answer is really up to us. What type of AI will we build? How will we solve the problem of ancestral memory? And what world will we leave for our children, both human and robot?

---

[7] *Hamlet*, Act II, Scene 2.

[8] "I'm fuzzy on the whole good/bad thing. What do you mean, *bad*?" — Venkman, *Ghostbusters*.

A

Distributions

A.1 Set Notation

We use the following conventions for standard sets:

$$\mathbb{Z} = \text{set of integers}$$
$$\mathbb{R} = \text{set of real numbers}$$
$$\mathbb{N} = \mathbb{Z}_* = \{n \in \mathbb{Z} : n \geq 0\}$$
$$\mathbb{Z}^+ = \{n \in \mathbb{Z} : n \geq 1\}$$
$$\mathbb{R}_* = \{x \in \mathbb{R} : x \geq 0\}$$
$$\mathbb{R}^+ = \{x \in \mathbb{R} : x > 0\}.$$

A.2 Simulation

With few exceptions, random samples are computed using the `np.random` package and distributions are computed using the `scipy.stats` package.

```
1  from numpy import random
2  from scipy import stats
3
4  samples = np.random.exponential(0.2, size=n)
```

Code Block A.1: Example usage of random samples and distributions

A.3 Discrete Distributions

A.4 Continuous Distributions

name	support	$p(x)$	$S(x)$	$\mathbb{E}[X]$	$\mathbb{V}(X)$	simulation
Bernoulli Bern(p)	$\{0,1\}$	$p^x(1-p)^{1-x}$	—	p	$p(1-p)$	random.binomial(1, p)
binomial Binom(n,p)	$\{0,\ldots,n\}$	$\binom{n}{x}p^x(1-p)^{n-x}$		np	$np(1-p)$	random.binomial(n,p) stats.binomial(n,p)
negative binomial NBD(r,p)	$\mathbb{N}$	$\dfrac{\Gamma(x+r)}{\Gamma(r)\Gamma(x+1)}p^r(1-p)^x$		$\dfrac{r(1-p)}{p}$	$\dfrac{r(1-p)}{p^2}$	random.negative_binomial(r,p) stats.nbinom(r,p)
shifted negative binomial sNBD(r,p)	$\{r, r+1,\ldots\}$	$\binom{x-1}{r-1}p^r(1-p)^{x-r}$		$\dfrac{r}{p}$	$\dfrac{r(1-p)}{p^2}$	random.negative_binomial(r,p)+r stats.nbinom(r,p,loc=r)
geometric Geom(p)	$\mathbb{N}$	$p(1-p)^x$	$(1-p)^{x+1}$	$\dfrac{1-p}{p}$	$\dfrac{1-p}{p^2}$	random.geometric(p)-1 stats.geom(p,loc=-1)
shifted geometric sGeom(p)	$\mathbb{Z}^+$	$p(1-p)^{x-1}$	$(1-p)^x$	$\dfrac{1}{p}$	$\dfrac{1-p}{p^2}$	random.geometric(p) stats.geom(p)

Table A.1: Catalogue of discrete distributions.

name	support	$p(x)$	$S(x)$	$\mathbb{E}[X]$	$\mathbb{V}(X)$	simulation
Poisson Poiss(λ)	$\mathbb{N}$	$\dfrac{e^{-\lambda}\lambda^x}{x!}$		λ	λ	random.poisson(lam) stats.poisson(lam)
beta-binomial BetaBin(n, α, β)	$\{0,\ldots,n\}$	$\dbinom{n}{x}\dfrac{B(\alpha+x, \beta+n-x)}{B(\alpha,\beta)}$	—	$\dfrac{n\alpha}{\alpha+\beta}$	$\dfrac{n\alpha\beta(\alpha+\beta+n)}{(\alpha+\beta)^2(\alpha+\beta+1)}$	N/A stats.betabinom(n,a,b)
beta-geometric BetaGeom(α, β)	$\mathbb{N}$	$\dfrac{B(\alpha+1, \beta+x)}{B(\alpha,\beta)}$	$\dfrac{B(\alpha, \beta+x+1)}{B(\alpha,\beta)}$	$\dfrac{\beta}{\alpha-1}$*	$\dfrac{\alpha\beta(\alpha+\beta-1)}{(\alpha-2)(\alpha-1)^2}$*	N/A N/A

Table A.2: Catalogue of discrete distributions (continued); * defined only when denominator is positive.

name	support	$p(x)$	$S(x)$	$\mathbb{E}[X]$	$\mathbb{V}(X)$	simulation
normal $N(\mu,\sigma^2)$	$\mathbb{R}$	$\dfrac{1}{\sigma\sqrt{2\pi}}e^{-(x-\mu)^2/(2\sigma^2)}$	$\Psi(x)$	μ	σ^2	`random.randn()*sig+mu` `stats.norm(loc=mu,scale=sig)`
chi-squared χ_p^2	$\mathbb{R}^+$	$\dfrac{x^{p/2-1}e^{-x/2}}{\Gamma(p/2)2^{p/2}}$		p	$2p$	`random.chisquare(p)` `stats.chi2(p)`
Student's T t_p	$\mathbb{R}$	$\dfrac{\Gamma((p+1)/2)}{\Gamma(p/2)\sqrt{p\pi}}\left(1+t^2/p\right)^{-(p+1)/2}$		0	$\dfrac{p}{p-2}{}^*$	`random.standard.t(p)` `stats.t(p)`
Snedecor's F $F_{p,q}$	$\mathbb{R}^+$	$\propto x^{p/2-2}\left[1+(p/q)x\right]^{-(p+q)/2}$		$\dfrac{q}{q-2}{}^*$		`random.f(p,q)` `stats.f(p,q)`
exponential $\mathrm{Exp}(\beta)$	$\mathbb{R}_*$	$\dfrac{1}{\beta}e^{-x/\beta}$	$e^{-x/\beta}$	β	β^2	`random.exponential(scale=b)` `stats.expon(scale=b)`

Table A.3: Catalogue of continuous distributions; * defined only when denominator is positive.

name	support	$p(x)$	$S(x)$	$\mathbb{E}[X]$	$\mathbb{V}(X)$	simulation
beta $\mathrm{Beta}(\alpha,\beta)$	$(0,1)$	$\dfrac{x^{\alpha-1}(1-x)^{\beta-1}}{B(\alpha,\beta)}$		$\dfrac{\alpha}{\alpha+\beta}$	$\dfrac{\alpha\beta}{(\alpha+\beta)^2(\alpha+\beta+1)}$	`random.beta(a,b)` `stats.beta(a,b)`
gamma $\mathrm{Gamma}(\alpha,\beta)$	$\mathbb{R}^+$	$\dfrac{x^{\alpha-1}e^{-x/\beta}}{\Gamma(\alpha)\beta^{\alpha}}$		$\alpha\beta$	$\alpha\beta^2$	`random.gamma(a,scale=b)` `stats.gamma(a,scale=b)`
Pareto $\mathrm{Pareto}(\alpha,\beta)$	$[\beta,\infty)$	$\dfrac{\alpha\beta^{\alpha}}{x^{\alpha+1}}$	$\left(\dfrac{\beta}{x}\right)^{\alpha}$	$\dfrac{\alpha\beta}{\alpha-1}$ *	$\dfrac{\alpha\beta^2}{(\alpha-1)^2(\alpha-2)}$ *	`random.pareto(a)*b+b` `stats.pareto(a,scale=b)`
Lomax $\mathrm{Lomax}(\alpha,\beta)$	$\mathbb{R}_*$	$\dfrac{\alpha}{\beta}\left[1+\dfrac{x}{\beta}\right]^{-(\alpha+1)}$	$\left[1+\dfrac{x}{\beta}\right]^{-\alpha}$	$\dfrac{\beta}{\alpha-1}$ *	$\dfrac{\alpha\beta^2}{(\alpha-1)^2(\alpha-2)}$ *	`random.pareto(a)*b` `stats.lomax(a,scale=b)`
Weibull $\mathrm{Weibull}(\alpha,\beta)$	$\mathbb{R}_*$	$\dfrac{\alpha}{\beta}\left(\dfrac{x}{\beta}\right)^{\alpha-1}e^{-(x/\beta)^{\alpha}}$	$e^{-(x/\beta)^{\alpha}}$	$\beta\Gamma(1+1/\alpha)$		`random.weibull(a)*b` `stats.weibull_min(a,scale=b)`

Table A.4: Catalogue of continuous distributions (continued).

References

Aalen, O.o., O. Borgan, H.K. Gjessing (2008) *Survival and Event History Analysis: A Process Point of View*, Springer.

Aggarwal, C.C. (2018) *Neural Networks and Deep Learning*, Springer.

Agresti, A. (2013) *Categorical Data Analysis*, 3rd ed., Wiley.

Agresti, A. (2015) *Foundations of Linear and Generalized Linear Models*, Wiley.

Agresti, A. (2019) *An Introduction to Categorical Data Analysis*, 3rd ed., Wiley.

Alpaydin, E. (2020) *Introduction to Machine Learning*, 4th ed., MIT Press.

Baldwin, R.R., W.E. Cantey, H. Maisel, and J.P. McDermott (1956) The optimum strategy in blackjack, *J. of the American Stat. Assoc.* **51**: 429–439.

Barber, D. (2012) *Bayesian Reasoning and Machine Learning*, Cambridge.

Beck, K., M. Beedle, A. van Bennekum, A. Cockburn, W. Cunningham, M. Fowler, J. Grenning, J. Highsmith, A. Hunt, R. Jeffries, J. Kern, B. Marick, R.C. Martin, S. Mellor, K. Schwaber, J. Sutherland, D. Thomas (2001) *The Agile Manifesto*, http://agilemanifesto.org.

Bentley, J.L. (1975) Multidimensional binary search trees used for associative search, *Communications of the ACM* **18**: 509–517.

Berger, P.D., R.E. Maurer, and G.B. Celli 2018 *Experimental Design: With Applications in Management, Engineering, and the Sciences*, 2nd ed., Springer.

Bergstra, J. and Y. Bengio (2012) Random search for hyperparameter optimization, *J. Machine Learning Research* **13**: 281–305.

Box, G.E.P., G.M. Jenkins, G.C. Reinsel, and G.M. Ljung (2016) *Time Series Analysis: Forecasting and Control*, 5th ed., Wiley.

Breiman, L. (1966) Bagging predictors, *Machine Learning* bf 26: 123–140.

Breiman, L. (2001) Random forests, *Machine Learning* bf 45: 5–32.

Brokwell, P.J. and R.A. Davis (1992) Time reversability, identifiability, and independence of innovations for stationary time series, *J. Time Series Analysis* **13**: 377–390.

Brokwell, P.J. and R.A. Davis (2016) *Introduction to Time Series and Forecasting*, 3rd ed., Springer.

Brown, J.W. and R.V. Churchill (2013) *Complex Variables and Applications*, 9th ed., McGraw–Hill.

Brier, G.W. (1950) Verification of forecasts expressed in terms of probability; *Monthly Weather Review* **78**(1)

Bühlmann, H. (1967) Experience rating and credibility, *ASTIN Bulletin* **4** 199–207.

Bühlmann, H. (1970) *Mathematical Methods in Risk Theory*, Springer.

Bühlmann, H. and A. Gisler (2005) *A Course in Credibility Theory and its Applications*, Springer.

Bühlmann, H. and E. Straub (1970) Glaubwürdigkeit für Schadensätze, *Bulletin of Swiss Ass. of Act.*, 111-133.

Buitinck, *et al.* (2013) API design for machine learning software: experiences from the scikit-learn project, in *ECML PKDD Workshop: Languages for Data Mining and Machine Learning*, p. 108–122.

Capinksi, M. and E. Kopp (2005) *Measure, Integral, and Probability*, 2nd ed., Springer.

Casella, G., and R.L. Berger (2002) *Statistical Inference*, 2nd ed., Brooks/-Cole.

Chawla, N.V., K.W. Bowyer, L.O. Hall and W.P. Kegelmeyer (2002) SMOTE: Synthetic Minority Over-sampling Technique, *J. Artificial Intelligence Research* **16**: p. 321–357.

Chen, T. and C. Guestrin (2016) XGBoost: A Scalable Tree Boosting System, *Proceedings of the 22nd ACM SIGKDD Int. Conf. on Knowledge Discovery and Data Mining* ACM 785-794.

Chollet, F. (2021) *Deep Learning with Python*, 2nd ed., Manning

Chong, E.K.P. and S.H. Zak (2008) *An Introduction to Optimization*, 3rd ed., John Wiley & Sons.

Colombo, R., W. Jiang (1999) A stochastic RFM model, *J. Interactive Marketing* **13**(3): 2–12.

Conway, J.B. (1978) *Functions of One Complex Variable I*, 2nd ed., Springer.

Cormon, T.H., C.E. Leiserson, R.L. Rivest, and C. Stein (2009) *Introduction to Algorithms*, 3rd ed., MIT Press.

Davidson-Pilon, C. (2016) *Bayesian Methods for Hackers: Probabilistic Programming and Bayesian Inference*, Addison Wesley. (See web for updated version `https://camdavidsonpilon.github.io/Probabilistic-Programming-and-Bayesian-Methods-for-Hackers/`.)

Devroye, L. (1986) *Non-Uniform Random Variate Generation*, Springer.

DiBenedetto, E. (2002) *Real Analysis*, Birkhäuser.

Dietterich, T. (2000) An experimental comparison of three methods for constructing ensembles of decision trees: bagging, boosting, and randomization, *Machine Learning* **40**: 139–158.

Dobson, A.J. and A.G. Barnett (2018) *An Introduction to Generalized Linear Models*, 4th ed., CRC Press.

Doerr, J. (2018) *Measure What Matters: How Google, Bono, and the Gates Foundation Rock the World with OKRs*, Portfolio.

Durbin, J. (1960) The fitting of time series models, *Review of the Institute of International Statistics* **28**: 233-244.

Dunn, P., and G.K. Smyth (2018) *Generalized Linear Models with Examples in R*, Springer.

Durrett, R. (2016) *Essentials of Stochastic Processes*, Springer.

Efron, B. (1971) Forcing a sequential experiment to be balanced, *Biometrika*, **58**, 403–417.

Efron, B. and T. Hastie (2016) *Computer Age Statistical Inference*, Cambridge.

Ekman, P. and W.V. Friesen (1971) Constants across cultures in the face and emotion, *J. Personality and Social Psych.* **17**(2): 124–129.

Ekman, P. (2007) *Emotions Revealed*, Holt.

Fader, P., B. Hardie, and K.L. Lee (2005) Counting your customers the easy way: an alternative to the Pareto/NBD Model. *Mark. Sci.* **24**: 275–284.

Fader, P.S., B.G.S. Hardie (2007) How to project customer retention, *J. Interactive Marketing* **21**(Winter): 76–90.

Fader, P., B. Hardie, J. Shang (2010) Customer-base analysis in a discrete-time noncontractual setting. *Mark. Sci.* **29**: 1086–1108.

Fisher, R.A. (1935) *The Design of Experiments*, Oliver and Boyd.

Flach, P. (2012) *Machine Learning: The Art and Science of Algorithms that Make Sense of Data*, Cambridge.

Folland, G.B. (1999) *Real Analysis: Modern Techniques and their Applications*, 2nd ed., Wiley.

Freund, Y. and R.E. Schapire (1996) Experiments with a new boosting algorithm, In *The Thirteenth International Conference on Machine Learning*, ed. L. Saitta, 148–156, San Mateo, CA, Morgan Kaufmann.

Greedy function approximation: a gradient boosting machine, *Annals of Stats.* **29**(5): 1189–1232.

Stochastic gradient boosting, *Comp. Stats. and Data Analysis* **38**(4): 367–378.

Gelman, A., J.B. Carlin, H.S. Stern, D.B. Dunson, A. Vehtari, and D.B. Rubin (2014) *Bayesian Data Analysis*, 3rd ed., CRC Press.

Gèron, A. (2019) *Hands-on Machine Learning with Sci-Kit Learn, Keras, and TensorFlow: Concepts, Tools and Techniques to Build Intelligent Systems*, 2nd ed., O'Reilly.

Gill, J. and M. Torres (2020) *Generalized Linear Models: A Unified Appraoch*, 2nd ed., Sage Publishing.

Goodfellow, I., J. Pouget-Abadie, M. Mirza, B. Xu, D. Warde-Farley, S. Ozair, A. Courville, Y. Bengio (2014) Generative Adversarial Nets, *Proc. of the Int. Conf. on Neural Information Processing Systems* (NIPS 2014): 2672–2680.

Goodfellow, I., Y. Bengio, and A. Courville (2016) *Deep Learning*, MIT Press.

Graesser, L. and W.L. Keng (2019) *Foundations of Deep Reinforcement Learning: Theory and Practice in Python*, Addison-Wesley.

Hajek, B. (2015) *Random Processes for Engineers*, Cambridge University Press.

Hamilton, J.D. (1994) *Time Series Analysis*, Princeton.

Härdle, W.K. and L. Simar (2019) *Applied Multivariate Statistical Analysis*, 5th ed., Springer.

Hastie, T, R. Tibshirani, and J. Friedman (2009) *The Elements of Statistical Learning: Data Mining, Inference, and Prediction*, Springer.

Hebb, D.O. (1949) *The Organization of Behavior: A Neuropsychological Theory*, Wiley.

Ho, T.K. (1995) Random decision forests, *Proc. 3rd Int. Conf. Document Analysis and Recognition*, Montreal, QC, 278–282.

Ho, T.K. (1998) The random subspace method for construction decision forests, *IEEE Transactions on Pattern Analysis and Machine Intelligence* **13**: 340–354.

Hochreiter, S. and J. Schmidhuber (1997) Long short-term memory, *Neural Computation* **9**(8): 1735–1780.

Hogg, R.V., E.A. Tanis, and D.L. Zimmerman (2015) *Probability and Statistical Inference*, 9th ed., Pearson.

Ishwaran, H., U.B. Kogalur, E.H. Blackstone, and M.S. Lauer (2008) Random survival forests, *Annals of App. Stats.* bf 2(3): 841–860.

Ishwaran, H. and U.B. Kogalur (2007) Random survival forests for R, *R News* bf 7(2): 25–31.

Imbens, G.W., and D.B. Rubin (2015) *Causal Inference for Statistics, Social, and Biomedical Sciences: An Introduction*, Cambridge University Press.

James, G., D. Witten, T. Hastie, and R. Tibshirani (2013) *An Introduction to Statistical Learning: With Applications in R*, Springer.

Johnson, M., M. Schuster, Q.V. Le, M. Krikun, Y. Wu, Z. Chen, N. Thorat, F. Viégas, M. Wattenberg, G. Corrado, M. Hughes, J. Dean (2016) Google's Multilingual Neural Machine Translation System: Enabling Zero-Shot Translation, `arXiv:1611.04558` [cs.CL].

Jung, C. (1916) The Structure of the Unconscious, reprinted in Jung, C. (1953) *Collected Works*, 263–292.

Jung, C. G. (2009). *The Red Book: Liber Novus*, edited by S. Shamdasani, W W Norton & Co.

Kahn, A.B. (1962) Topological sorting of large networks, *Communications of the ACM*, **5**(11): 558–562.

Kiefer, J. (1953) Sequential minimax search for a maximum, *Proceedings of the American Mathematical Society* **4**(3): 502–506.

Klein, J.P. and M.L. Moeschberger (2003) *Survival Analysis: Techniques for Censored and Truncated Data*, 2nd ed., Springer.

Kochenderfer, M.J. (2015) *Decision Making Under Uncertainty*, MIT Press.

Kohavi, R., D. Tang, and Y. Xu (2020) *Trustworthy Online Controlled Experiments: A Practical Guide to A/B Testing*, Cambridge University Press.

Kuhn, M. and K. Johnson (2013) *Applied Predictive Modeling*, Springer.

Lafore, R. (2003) *Data Structures and Algorithms in Java*, 2nd ed., SAMS.

Lambert, K.A. (2014) *Fundamentals of Python Data Structures*, Cengage.

Lee, K.D. and S. Hubbard (2015) *Data Structures and Algorithms with Python*, Springer.

Maruskin, J. (2010) Distance in the space of energetically bounded Keplerian orbits, *Celestial Mechanics and Dynamical Astronomy* **108**: 265–274.

Maruskin, J.M., D.J. Scheeres, and K.T. Alfriend (2012) Correlation of optical observations of objects in earth orbit, *Journal of Guidance Control and Dynamics* **32**: 194–209.

Maruskin, J. (2018) *Dynamical Systems and Geometric Mechanics: An Introduction*, 2nd ed., de Gruyter.

McCulloch, W.S. and W. Pitts (1943) A logical calculus of the ideas immanent in nervous activity, *The Bulletin of Mathematical Biophysics* **5**: 115–133.

McElreath, R. (2016) *Statistical Rethinking: A Bayesian Course with Examples in R and Stan*, CRC Press.

Metropolis, N. (1987) The beginning of the Monte Carlo method, *Los Alamos Science* (1987 Special Issue dedicated to Stanislaw Ulam): 125–130.

Mischel, W. and E.B: Ebbesen (1970) Attention in delay of gratification, *Journal of Personality and Social Psychology* **16**(2): 329–337.

Mood, A.M. (1950) *Introduction to the Theory of Statistics.* McGraw-Hill.

Morgan, S.L. and C. Winship (2015) *Counterfactuals and Causal Inference: Methods and Principles for Social Research*, 2nd ed., Cambridge University Press.

McKinney, W. (2017) *Python for Data Analysis*, 2nd ed., O'Reilly.

Mnih, V., K. Kavukcuoglu, D. Silver, A. Graves, I. Antonoglou, D. Wierstra, M. Riedmiller (2013) Playing Atari with Deep Reinforcement Learning, arXiv:1312.5602 [cs.LG].

Newbold, P. (1974) The exact likelihood function for a mixed autoregressive moving average process, *Biometrika* **61**: 423–426.

Niculescu-Mizil and Caruana (2005) Predicting good probabilities with supervised learning, In ICML-05 *International Conference on Machine Learning*, Aug. 2005, 625–632; https://doi.org/10.1145/1102351.1102430.

Nielson, A. (2019) *Practical Time Series Analysis* O'Reilly.

Nyce, C.M. (2017) The winter getaway that turned the world upside down, *The Atlantic*.

Olive, D. (2017) *Linear Regression*, Springer.

van den Oord, A., S. Dieleman, H. Zen, K. Simonyan, O. Vinyals, A. Graves, N. Kalchbrenner, A. Senior, K. Kavukcuoglu (2016) WaveNet: A Generative Model for Raw Audio, arXiv:1609.03499 [cs.SD].

Pearl, J. (2009) *Causality: Models, Reasoning, and Inference*, 2nd ed., Cambridge University Press.

Pearl, J., M. Glymour, and N.P. Jewell (2016) *Causal Inference in Statistics: A Primer*, Wiley.

Pedregosa, *et al.* (2011) Scikit-learn: Machine Learning in Python, *Journal of Machine Learning Research* **12**: p. 2825–2830.

Platt, J. (2000) Probabilistic outputs for support vector machines and comparison to regularized likelihood models; In Bartlett B., B. Schölkopf, D. Schuurmans, A. Smola (eds.) *Advances in Kernel Methods Support Vector Learning*, p. 61–74, MIT Press.

Reinhart, A. (2015) *Statistics Done Wrong*, no starch press.

Rizzolatti, G., L. Fadiga, V. Gallese, and L. Fogassi (1996) Premotor cortex and the recognition of motor actions, *Cognitive Brain Research* **3**(2): 131–141.

Rizzolatti, G. and C. Sinigaglia (2008) *Mirrors in the Brain: How We Share our Actions and Emotions*, Oxford University Press.

Robbins, H. (1952) Some aspects of the sequential design of experiments, *Bulletin of the American Mathematical Society* **58**(5): 528–535.

Rosenbaum, P.R. (2002) *Observational Studies*, 2nd ed., Springer.

Rosenbaum, P.R. (2019) *Observational and Experiment: An Introduction to Causal Inference*, reprint ed., Harvard University Press.

Rosenbaum, P.R. (2020) *Design of Observational Studies*, 2nd ed., Springer.

Rosenblatt, F. (1957) The Perceptron: a perceiving and recognizing automation, *Cornell Aeronautical Laboratory Reprt*

Ross, S. (2012) *A First Course in Probability*, 9th ed., Prentice Hall.

Salvatier, J., T.V. Wiecki, and C. Fonnesbeck (2016) Probabilistic programming in Python using PyMC3, *PeerJ Computer Science* 2:e55 DOI: 10.7717/peerj-cs.55.

Schapire, R.E. (1990) The strength of weak learnability, *Machine Learning* **5**: 197–227.

Schapire, R.E., Y. Freund, P. Bartlett, and W.S. Lee (1998) Boosting the margin: a new explanation for the effectiveness of voting methods, *Information Sciences* **179**: 1298–1318.

Schmittlein, D.C., D.G. Morrison, R. Colombo (1987) Counting your customers: who are they and what will they do next? *Manag. Sci.* **33**-1.

Schmittlein, D.C. and R.A. Peterson (1994) Customer base analysis: An industrial purchase process application. *Marketing Sci.* **13**(1): 41–67.

Schuster, M., M. Johnson, and N. Thorat (2016) Zero-Shot Translation with Google's Multilingual Neural Machine Translation System, *Google Brain Team and Google Translate*, via `ai.googleblog.com`.

Seber, G.A.F. and A.J. Lee (2003) *Linear Regression Analysis* (2nd ed.), John Wiley.

Selvamuthu, D. and D. Das (2018) *Introduction to Statistical Methods, Design of Experiments, and Statistical Quality Control*, Springer.

Selvin, S. (1975a). A problem in probability (letter to the editor), *The American Statistician* **29**(1): 67–71.

Selvin, S. (1975b) On the Monty Hall problem (letter to the editor), *The American Statistician* **29**(3): 134.

Seufert, E.B. (2014) *Freemium Economics: Leveraging Analytics and User Segmentation to Drive Revenue*, Morgan Kaufmann, an imprint of Elsevier.

Shao, J. (2003) *Mathematical Statistics*, 2nd ed., Springer.

Shumway, R.H. and D.S. Stoffer (1999) *Time Series Analysis and Its Applications: With R Examples*, 4th ed., Springer.

Silver, D., *et al.* (Google Deep Mind) (2016) Mastering the game of Go with deep neural networks and tree search, *Nature* **529** 484–489.

Stevens, E., L. Antiga, and T. Viehmann (2020) *Deep Learning with PyTorch*, Manning.

Sutton, R.S. and A.G. Barto (2018) *Reinforcement Learning: An Introduction*, 2nd ed., MIT Press.

Takeuchi, H. and I. Nonaka (1986) The new new product development game, *Harvard Business Review*

Thorp, E.O. (1966) *Beat the Dealer: A Winning Strategy for the Game of Twenty-One*, Revised ed., Vintage.

Thorp, E.O. (2017) *A Man for all Markets: From Las Vegas to Wall Street, how I beat the dealer and the market*, Random House.

Trussell, J. and D.E. Bloom (1979) A model distribution of height or weight at a given age, *Human Biology* **51**: 523-536.

Turing, A.M. (1950) Computing machinery and intelligence, *Mind* **59**(236): 433–460.

Vaupel, J.W. and A.I. Yashin (1985). Heterogeneity's ruses: some surprising effects of selection on population dynamics, *The American Statistician* **39**: 176–185.

Wasserman, L. (2004) *All of Statistics: A Concise Course in Statistical Inference*. Springer.

Wasserman, L. (2006) *All of Nonparametric Statistics*. Springer.

Watkins, C.J.C.H. (1989) *Learning from Delayed Rewards*, PhD thesis, University of Cambridge.

Watkins, C.J.C.H. and P. Dayan (1992) Q-learning, *Machine Learning* **8**:279–292.

Wei, W.W.S. (2019) *Time Series Analysis: Univariate and Multivariate Models*, Pearson.

Withers, C.S. and S. Nadarajah (2014) The spectral decomposition and inverse of multinomial and negative multinomial covariances. *Brazilian Journal of Probability and Statistics*. Vol. 28, No.3, p. 376–380.

Wolfram Alpha LLC (2018) *Wolfram—Alpha*, wolframalpha.com.

Zou, H., T. Hastie, and R. Tibshirani (2007) On the "degrees of freedom" of the lasso, *The Annals of Statistics* **35**(5): 2173–2192.

Index

www.ingramcontent.com/pod-product-compliance
Lightning Source LLC
Chambersburg PA
CBHW071329210326
41597CB00015B/1382